ALSO BY CHRISTOPHER McGOWAN

The Successful Dragons:
A Natural History of Mesozoic Reptiles

In the Beginning:
A Scientist Shows Why the Creationists Are Wrong

Dinosaurs, Spitfires, and Sea Dragons

Discover Dinosaurs

Diatoms to Dinosaurs:
The Size and Scale of Living Things

Make Your Own Dinosaur out of Chicken Bones

The Raptor and the Lamb

CHRISTOPHER
McGOWAN

HENRY HOLT AND COMPANY
NEW YORK

The Raptor AND THE Lamb

Predators and Prey in the Living World

Henry Holt and Company, Inc.
Publishers since 1866
115 West 18th Street
New York, New York 10011

Henry Holt® is a registered trademark of Henry Holt and Company, Inc.

Published in Canada by Fitzhenry & Whiteside, Ltd.
195 Allstate Parkway, Markham, Ontario L3R 4T8.

Library of Congress Cataloging-in-Publication Data
McGowan, Christopher.
The raptor and the lamb: predators and prey in the living world/
Christopher McGowan.—1st ed.
p. cm.
Includes bibliographical references and index.
1. Predation (Biology) I. Title.
QL758.M35 1997 96-53506
CIP

ISBN 0-8050-4298-9

First Edition 1997

Designed by Kate Nichols
Illustrations by Abigail Rorer

Printed in the United States of America
All first editions are printed on acid-free paper. ∞

1 2 3 4 5 6 7 8 9 10

To Liz with love:

Wife. Best friend. Terra firma.

Contents

Acknowledgments

First and foremost I wish to thank the authors of the scientific publications upon which this book is based. I thank them for their curiosity, their industry, and for their ingenuity in unraveling the complexity of living things. There is a joy in understanding that transcends economic necessity; let us continue to explore. Some of these researchers—all busy people and mostly unknown to me personally—generously took time to critically review individual chapters in their areas of expertise. I am deeply moved by their interest and support, and for their valuable advice for improvements. I received references, reprints, pre-prints, unpublished data, and answers to various questions by correspondence. I also received warm encouragement when most it was needed. How can I ever thank you David Bird, Marie Bundy, Sarah Durant, Malcolm Edmunds, Howard Evans, Brock Fenton, Lyn Forster, David Gaskin, Frank Götmark, Peter Herring, Christine Janis, Kenneth Kardong, Peter Klimley, Zoe Lucas, John Moore, Ian Newton, Kevin Padian, Gary Polis, John Rubin, Bev Scott, Mike Taylor, Timothy Tricas, Blaire Van Valkenburgh, David Weishampel, Peter Wellnhofer, Hans Winter, and Jeannette Yen?

Abigail Rorer's gift for distilling the essence of the animate world in her drawings is self-evident from the pages that follow. It has been a privilege and joy working with such a professional, and I thank her for the beauty that graces this book.

Allen Peacock, my editor, was enthusiastic about the project from the outset. I thank him for his constancy, and for all those things that editors do to see a manuscript through to fruition. My thanks also to Martine Bellen who has assisted in all these things. My sincere thanks to Jeanne Tift, also of Holt, for her meticulous care in editing the manuscript—your attention to detail is much appreciated. I also thank Kate Nichols and Paula Szafranski for the book design, and Eva Diaz, the production manager, for shepherding the manuscript through all its stages. Thanks also to Vicki Haire and Chris Potash for their careful copyediting of the final manuscript.

The Royal Ontario Museum houses extensive collections of most of the major groups of animals, and I am very fortunate to work in such a place. I am equally fortunate in having colleagues with whom I can discuss the biology of these organisms. I am grateful to you all for your help, with particular thanks to Allan Baker, Dale Calder, Douglas Currie, Chris Darling, Bob Murphy, Glenn Murphy, Mark Peck, Marty Rouse, and Rick Winterbottom. My sincere thanks to Catherine Skrabec and Michele Bobyn for their assistance in obtaining reference material. I also want to thank the staff of the Library, Royal Ontario Museum, for assistance with inter-library loans.

I gratefully acknowledge my faculty appointment to the Department of Zoology, University of Toronto, and to generations of fine students who have made me strive to understand more of the living world. My thanks also to the Natural Sciences and Engineering Research Council of Canada which has supported my research for over two decades.

How very fortunate I am to be represented in my writing endeavors by an agency, Scovil, Chickak and Galen, that takes such good care of its clients. I thank Shawna McCarthy and Russ Galen for helping me take the initial steps, and Jill Grinberg for helping me the rest of the way. Jill has been all, and more, that anyone could ever hope for in an agent. During the early stages of this book—markedly different from all my others—I was so unsure of the merits of what I had produced that the whole project was in jeopardy of stalling. Jill took the time to read what I had written, and the encouragement and reassurance she gave rocketed me into the blue. No words can express my gratitude.

Work on this book began during an unforgettable snowy weekend, marooned in the foothills of the Black Mountains in Wales. And what better place to begin writing than an antiquarian book shop in the legendary book kingdom of Hay-on-Wye? Warm thanks to my good friends Chris and Cathy Arden, owners of the finest natural history book shop in Hay.

Leaving the best to last, I thank Liz, my wife, for all the love and support she gives. In spite of her own busy schedule she still makes time to read early drafts, proofread galleys, give sound advice, and keep me sailing on an even keel. What would I do without you?

The Raptor and the Lamb

Prologue

Most animals are either eaten or eat other animals. Plants, too, are often consumed by animals. Consequently the chances of being devoured, or of eating some other organism in order to survive, are exceedingly high. The evolutionary pressures to succeed as a predator, or to succeed at not becoming a prey, are accordingly high. So too are the pressures to succeed as a herbivore, or, from the plant's perspective, to succeed at not being destroyed by browsers and grazers. The evolutionary struggle that pits predator against prey, or plant against herbivore, is often likened to an arms race, but the analogy, which oversimplifies the situation, is not a perfect one. If the antelope evolved higher running speeds to escape the lion, which in turn became more fleet-footed to catch the antelope, which then became even faster to escape the lion—ad infinitum—they would both become the fastest animals on Earth. This competition absurdum, which characterized the arms race between the superpowers, is not the hallmark of the rest of the living world. There are two main reasons why such an evolutionary escalation cannot work in nature. First, killing, or avoiding being killed, is not the only item on an organism's agenda, and an organism cannot be optimally evolved for all things. Just as an aircraft cannot be designed to carry heavy loads and to fly at the fastest speed, so a predator cannot be both the fastest and the biggest and strongest. The cheetah, for example, is the fastest land animal, reaching speeds of up to about

70 miles per hour (mph), or 110 kilometers per hour (km/h). Its superior speed gives the cheetah a much higher hunting success over slower predators, like the lion, whose top speed is less than 40 mph (65 km/h). But cheetahs pay a high price for their speed: their smaller size, and more slender build, make them vulnerable to attack from lions, which often rob them of their hard-earned quarry. Lions are also able to tackle much larger animals than cheetahs, giving them a larger choice of potential prey. Similar compromises occur among prey species. The long slender legs of an antelope give it a higher top speed than a lion, but its smaller leg and back muscles give it an inferior acceleration. The acceleration of the antelope would be improved if its body were lighter. This could be achieved by having a smaller gut, but then its digestive efficiency would be compromised.

It has been suggested that there should be higher evolutionary pressures on prey species than on predators because if a predator loses a contest with a potential prey, it only loses a meal, whereas the other loses its life. This "life-dinner" argument may sound plausible, but it does not stand up to close inspection. For if a predator loses too many meals, it will be just as dead as a lamb in the jaws of a wolf. By the same token, if a potential prey animal spent all its energies fleeing from predators, it would be unable to feed properly and would similarly be doomed. Predator and prey alike have evolved strategies to improve their respective chances of success, but not at the exclusion of all else. The same is true for plants and the animals that would destroy them. Nor is it necessarily an advantage for one of the antagonists to respond to improvements in the other. Suppose a prey species evolved some improvement in its defensive strategies, perhaps better acceleration performance or more alertness to potential dangers. If the predator did not coevolve some corresponding improvement in its hunting techniques, it would catch fewer prey. The number of prey would therefore increase, but this could improve the predator's chances of catching them, so, in the long term, its hunting success might not suffer.

The situation between predator and prey may be thought of in terms of risk management. When a commercial airliner lifts off from the Tarmac, it often has a list of minor defects that need attention. None are serious enough to endanger the lives of the passengers, but some may reduce the safety of the flight by some infinitesimally small amount. The risk factor would be entirely eliminated if the aircraft were serviced each time a problem developed, but that would be too expensive in terms of downtime. The airline therefore manages the risk by waiting until a number of items need attention before taking the aircraft out of service for maintenance. When a hungry lion approaches a herd of wildebeest, each individual has a very small chance of becoming its prey. But the risk does not justify the evolution of antipredatory strategies over and above those already possessed by the species. Indeed, the risk of being eaten does not warrant an individual stopping its grazing, at least not until an attack is imminent. The costs of evolving antipredatory adaptations within a species are therefore balanced against the risks of being killed.

Another reason why the arms-race analogy is an oversimplification of the living world is that predators usually target many different prey species. Similarly herbivores, from caterpillars to elephants, usually attack more than one species of plant. And the response of one antagonist toward another is determined by the species involved. A herd of zebra, for example, might allow a cheetah to approach within twenty yards before fleeing, whereas a lion might not be allowed to get any closer than forty yards. This is because there is a greater likelihood of a lion attacking a zebra than of a cheetah, because of the differences in size and strength between the two predators.

Although being eaten by a predator is disastrous for the individual, the action is not usually detrimental to the population as a whole. Indeed, the continual cropping of prey species by predators is often to their collective benefit. For when animals are left without the constraint on their population growth provided by predators, many of them will exhaust their food supply and suffer the consequences of starvation. Predation by wolves, for example, has saved moose from eating themselves out of food on an island in Lake Superior where food resources are limited. Predator and prey species are therefore to some extent mutually beneficial, helping to maintain the dynamic balance between population sizes and natural resources. Plants have a somewhat different mutualism with certain animals, especially insects, even though these animals may be destructive. Caterpillars, for example, can cause considerable damage to plants, but adult butterflies, like bees and many other flying insects, play a vital role in cross-pollination, thereby contributing to the plants' survival.

Some predator-prey interactions are characterized by remarkably complex adaptations, especially when these interactions involve only two species. There is a species of spider, for example, that lures the male of a particular moth species to its doom through the following deception: the spider releases a scent into the air that mimics the sex hormone used by the female moth to attract mates. Explaining how such an elaborate ruse may have evolved is intellectually challenging. A whole book might have been written on the possible evolutionary pathways leading to such predator-prey interactions, but much of it would have been speculative. My primary interest in this book is rather to look at how living organisms are adapted to their lifestyles. The question of how certain features may be correlated with certain functions is a more straightforward one to address, but we certainly do not have all the answers—far from it. I have been studying animals and their complex ways far too long to believe in "just so" stories.

Let me give just one example to illustrate my point. When I last visited the Metro Toronto Zoo, one of the staff members, who was showing me around, pointed to an enclosure that housed a wildebeest. These cow-sized herbivores have sharp handlebar horns, and my host commented on how dangerous the animal was. The keepers, mindful of how much damage a wildebeest can inflict with its horns, avoid going into the enclosure if they possibly can. However, a wildebeest in the wild will often stand by and watch a lion kill its calf without making any attempt to attack it. This paradox can be rationalized in terms of the primary function of horns in ritualized fighting between rival males, but I still find the explanation less than satisfactory.

Phylogeny, the study of relationships among organisms, has made considerable progress during the last quarter century. Part of this progress is attributable to the availability of new techniques for assessing such relationships. One of the most powerful of these tools is DNA analysis, where the triplet codes contained in the genes are revealed in the laboratory. DNA sequencing, for example, has revealed that our closest relative among the primates is the chimpanzee rather than any of the other great apes. The objective of phylogenetics is to establish natural groupings of organisms, that is, a group of organisms that evolved from the same, common, ancestor. Insects, for instance, form a natural, or *monophyletic*, group, and it is believed that they all evolved from a common ancestor. Descent from a common ancestor is established by looking for specialized, or derived, features that all members of the group share. Insects share a number of derived features, mostly pertaining to details of their internal anatomy. These include the possession of an ovipositor, a structure used for depositing eggs, and valves on the holes (spiracles) through which insects breathe. The evolutionary pathways that a group of organisms can follow are in part determined by their phylogeny. Dogs, for example, as discussed in the first chapter, are unable to capture their prey by seizing them in their paws (as cats do) because of the restricted wrist movements they inherited from their ancestors. The constraints placed upon organisms by the phylogenetic baggage they carry should be borne in mind in the pages that follow.

We have a fatal attraction for predators, possibly because we often fell prey to their attentions during more remote periods in our history. We also have a morbid curiosity for the fate of the quarry—how the last desperate moments of their lives are played out. But not all encounters between hunter and hunted are as blatantly malevolent as the wolf's killing of the lamb. Far more common is a subtle struggle of offensive and counteroffensive strategies, reminiscent of the wars we humans wage among ourselves.

I was prompted to write this book by something I wrote in my *Diatoms to Dinosaurs* about flying insects. The section in question concerned the subtle defense strategies of some moths. Most moths are nocturnal and are heavily preyed upon by bats, which are also creatures of the night. Bats are such highly maneuverable fliers that they are more than a match for moths. Furthermore, they have evolved a sophisticated sonar system for detecting and homing in on their prey. This advanced guidance system, coupled with their superior flying performance, makes them the most formidable moth predators. Many moths have evolved sensitive hearing, enabling them to detect the high-pitched sonar signals of approaching bats. Some moths respond to the sounds of bats by diving to the ground. Some others, remarkably, have evolved a sonar-jamming system and emit high-frequency sounds that interfere with the sounds produced by the bats. This singular example of the coevolution of offensive and defensive strategies captured my imagination, spawning the idea of the present book.

The coevolution of offensive and defensive weapon systems is a fascinating area of research that has attracted a considerable amount of attention. There is consequently an extensive literature on the subject, but it tends to be widely dispersed among the scientific journals, with few popular accounts. I wrote this book with a general reader in mind. By that I mean a reader who is interested in natural history but who does not necessarily have a scientific background. However, I hope that some of my colleagues and students will also read the book and get something out of it. I did not write the book as a text, and the extensive scientific literature upon which it is based is therefore not referenced throughout in the usual way. Instead, I provide references to the pertinent works for each of the chapters at the end of the book. The subject matter covers a broad spectrum of organisms, from acacia trees and potato plants to wasps and killer whales. Trying to organize such a rich diversity of topics was a challenge in itself, and any subdivision of the material is bound to have some shortcomings. I structured each chapter around a particular group of organism, such as warm-blooded land animals or microscopic animals that live in the sea. Each chapter essentially stands alone, but there are common patterns and themes running throughout. I hope that by book's end, an overall picture of the unity of the living world will have emerged. I purposely include many examples of different species in all of the chapters to convey something of the richness of the biodiversity that still remains on our overcrowded planet.

The book begins on the African plains, with some of the more familiar examples of predator-prey interactions. The players in these acts are all land mammals. Chapter 2 turns the spotlight on reptiles, whose fundamentally different physiology from that of mammals causes them to adopt entirely different hunting strategies. The third chapter is devoted to vertebrates living in the sea. Water is almost one thousand times denser than air, and this factor regulates the pace of predatory encounters. Water is also buoyant, permitting marine animals to attain sizes unsurpassed on land. As a respite from the struggles between hunter and hunted, the fourth chapter deals with camouflage and how animals make themselves less conspicuous. The fifth chapter takes up the struggle again, this time in the air; here some of the encounters between predators and their prey take place at breakneck speeds. Chapter 6 is devoted to some of our worst nightmares—spiders, scorpions, and other joint-legged (arthropod) animals. Arthropod hunting strategies range from the unsophisticated pugilism of scorpions, to the elaborate snares of lone spiders, to the coordinated attacks of armies of ants. A trip back to the remote past of the Age of Dinosaurs in chapter 7 gives flight to fancy. But how accurately do the popular images of dinosaurs and their kin depict life on Earth over 65 million years ago? Although the world back then was quite unlike our modern times, some of the players had similar roles to play. Chapter 8 deals with the minutia of life—the plankton—whose members are as beautiful and bizarre as some of our most imaginative dreams. Here life is dominated by fluid forces that make the organisms appear to be swimming in thick syrup. Some planktonic predators exploit the consequences of this novel situation to detect their prey through remote sensing. Chapter 9, the last chapter, is concerned with plants and their interminable struggle with animals. Although some animals, notably insects, wreak havoc upon them, plants are dependent upon animals for cross-pollination. The book ends in an English country garden, that time-honored bastion of peace and tranquillity.

One last note. It has not been my intention to indulge in gratuitous violence—goodness knows we see and hear far too much of that in our everyday lives. Nevertheless, much of the subject matter in this book is violent by its very nature and is therefore disturbing. I was disturbed by some of the things I learned and wrote of here. Seeing an animal eaten alive is intensely distressing, but it is a normal part of nature. We must avoid viewing these acts of naked aggression in human terms. As far as we know, only the human animal has a conscience, has morals. It is therefore irrational to look upon any of the seemingly cold-blooded killings in the pages that follow as cruel. Nor should we make value judgments against predators in favor of their prey. It is just as much a part of the normal scheme of things for a lion to snuff out the life of an antelope as it is for the antelope to graze upon the grass.

1

Nature, Red in Tooth and Claw

The moon, in its last quarter, spills its cold brittle light across the black African landscape. A small pride of lions—four lionesses and their shaggy-maned lord—doze fitfully beside a low thornbush. Waiting. A marabou stork, all legs and beak and smelling of offal, stretches a wing and goes back to sleep with a rustle of feathers. Unseen creatures add their chirping, laughing, shrieking voices to the night song of insects. Way off in the distance is a small herd of zebras. They stand in a huddle, half asleep, half alert, big eyes bulging into the night. There are no thickets or trees to conceal a would-be predator, so they are not unduly concerned for their safety.

The early evening sky had been ablaze with stars, but clouds drifted in soon after midnight, blocking out large patches of the heavens. One of the lionesses gets to her feet, stretches, yawns, and lets out a low moaning grunt. It is time. The other lionesses are by her side in a moment, but the male, though wide awake, makes no attempt to rise. The four females amble off in single file, their pace unhurried. A few moments pass, then the male gets to his feet and reluctantly shuffles off in the same direction, keeping well behind them. An ethereal wisp of cloud drifts across a horn of the moon.

The lionesses catch sight of the zebras from a distance of about half a mile. Eight pairs of eyes lock onto their targets as the predators spread out in a ragged line abreast. They quicken their pace to a brisk walk. Lions do most of their hunting at night, and, like most other cats (felids), they have excellent night vision. Their modified eye has a large pupil and lens, to gather as much light as possible.[1]

The lionesses, having closed the gap to two hundred yards, begin stalking, each one crouching down low in a sinuous crawl through the bone-dry grass. One takes the lead, the other three dropping back and fanning out in a flanking maneuver. Veterans of the chase, they know the importance of concealment and sneak only the occasional glimpses of their quarry across the top of their cover. The chances of being seen are considerably reduced by the darkness—the success of communal hunting is almost doubled at night. But the crescent moon still reflects enough light to form silhouettes against the vastness of the sky, and the big cats take no chances. The male lion, who had dropped back some sixty yards behind the others, walks along cautiously, not taking his eyes off the herd. A zebra stamps a hoof against the hard ground, causing all four lionesses to freeze in midstride. Some of the other zebras shuffle and snort before returning to their uneasy rest. The predators slink closer.

A lion (*Panthera leo*) can run at speeds of over 30 mph (49 km/h), but zebras (*Equus burchelli* and *E. grevyi*), like most of their prey, can run faster, exceeding 40 mph (65 km/h). However, the lion, with its muscular limbs and powerful body, has the superior acceleration. If it can get within 50 yards of its prey, preferably less, it has a reasonable chance—a chance to close the gap in an explosive rush before the other has accelerated to its superior speed.

The leading lioness has reduced the distance to 100 yards (92 m). Nearly there. Another cloud passes across the moon, but this time it is a large enough to snuff out the lunar light like a flame. The lioness responds instantly, throwing herself toward the unseeing zebras with unbridled power. The zebras, sensing imminent danger, bunch up and begin a wheeling maneuver, hooves flying, dust rising. The big cat selects her quarry—an instantaneous decision based on the zebra's proximity and reaction. It was not the closest one, but it failed to follow the choreography of the herd quite as closely as the others. Locking onto her target, the lioness follows its every move, rapidly closing the distance between them.

Zebras weigh about twice as much as lions, but when a lioness traveling at 30 mph (49 km/h) launches her 280 pounds (130 kg) of muscle and bone at the haunches of a fleeing zebra, the energy transfer is sufficient to throw the prey off balance, sending it in a crashing heap to the ground. Lions, like domestic cats, sheath their claws to keep them sharp.

The lioness sinks her scimitar talons into the zebra's rump. They rip through the tough hide and anchor deep in the muscle. The startled animal lets out a loud bellow as its body hits the ground. An instant later the lioness releases her claws from its buttocks and sinks her teeth into the zebra's throat, choking off the sound of terror. Her canine teeth are long and sharp, but an animal as large as a zebra has a massive neck, with a thick layer of muscle beneath the skin, so although the teeth puncture the hide they are too short to reach any major blood vessels. She must therefore kill the zebra by asphyxiation, clamping her powerful jaws around its trachea (windpipe), cutting off the air to its lungs. It is a slow death. If this had been a small animal, say a Thomson's gazelle (Gazella thomsoni) the size of a large dog, she would have bitten it through the nape of the neck; her canine teeth would then have probably crushed the vertebrae or the base of the skull, causing instant death. As it is, the zebra's death throes will last five or six minutes. The lioness, lying on her haunches, has positioned herself in front of the zebra so as to be out of harm's way of flaying hooves, but her captive lies remarkably still, as if accepting the inevitable—it is probably in a state of shock. Its eyes bulge wildly from their sockets, while its tongue, black in the returning moonlight, lolls in the dust. And as the huntress patiently applies the life-robbing pressure with her jaws, the other three lionesses join her. One of them settles beside the zebra's haunches and licks at the blood trickling from the claw wounds. She then sinks her teeth into the leg and rips away at the living muscle. With much snarling and jockeying for position, she is joined by the other two lionesses.

The three lionesses have barely begun feeding when the male arrives. Although

Lions do much of their hunting at night.

he has played no part in the hunt, he roars imperiously, and they relinquish their positions to let him feed. Growling and hissing at his consorts, he sinks down beside the zebra. He does nothing for several long moments, as if savoring his place at the high table. Then he laps at the blood flowing from the wound in the zebra's haunches. Slowly at first, then greedily. He nips off a morsel of flesh with his incisors, snarling malevolently. Then he begins feeding in earnest. Meanwhile the first lioness continues her death grip on the zebra's throat. A hoof begins twitching uncontrollably.

One of the displaced lionesses rips open the zebra's abdomen and begins feeding on the warm viscera. She is joined by the other lionesses. It is possible that lions eat parts of the gut and other abdominal organs to satisfy their need for vitamins and other essential nutrients that they cannot obtain from flesh. Blood flows freely from a number of major vessels that have been severed, filling the abdominal cavity and spilling onto the ground. With a final convulsion of its body, the zebra dies. Nature, red in tooth and claw.

This last line, from Tennyson, captures the popular notion of the interminable struggle for survival among species—the driving force leading to the appearance of new forms—but Darwin saw it as a more subtle process. His five-year trip around the world aboard the *Beagle* (1831–36) had persuaded him that present-day species had arisen—evolved—from previously existing species, rather than having been created by the hand of God as told in the Scriptures. He spent the next two decades searching for a likely mechanism. He was impressed by the changes that selective breeding had brought about during the domestication of animals over the relatively short time of a few centuries. Just consider the great variety of dogs or sheep or cattle we have today, all the result of selective breeding. If humans could produce such diversity by selective breeding in a millennium or less, nature could surely do better over millions of years. Darwin's problem was to deduce how.

He recognized that more individuals of a given species are born than can possibly survive. He also acknowledged that some individuals inherit favorable traits from their parents that can improve their chances of survival. Putting these two premises together gave him a mechanism for evolution. Animals possessing inheritable traits that give them a better chance of surviving will pass these features on to their offspring, giving them a better chance too. The accumulation of these favorable features over countless thousands of years would bring about a modification of the species, causing it to become better adapted to its environment. He termed this mechanism—the selection of advantaged individuals—*natural selection.* Natural selection, in concert with environmental change, eventually gives rise to new species. The driving force for change is called *selection pressure.* The zebras on the African plain, for example, are under selection pressure from predators, among other things. The zebra that was killed by the lioness was the one that failed to keep together with the rest of the herd. The zebra's behavior may have been attributed to a number of factors, ranging from its running abilities to its spatial awareness. If the behavior had genetic components, the zebra's unfavorable genes would have been removed, by natural selection, from the *gene pool* of the herd, which is the sum total of genes in that breeding population.

Sunrise, and the five lions have eaten their fill of the zebra, snacking and dozing through the small hours. Their abdomens are grossly distended. The lionesses are stretched out on the ground like revelers after a party, but the male is on his feet beside the kill. The carcass, stripped of flesh and much of the hide, has been partially dismembered. The remains of one hind leg lie beside one of the lionesses; the other lies abandoned in the dust. The ends of the ribs have been gnawed; so too have the edges of the shoulder blades and the bones and cartilage of the nasal region. About the only portion of the soft parts they have not devoured is the large intestine, with its contents of partially digested grass.

*The pride is not alone. A small group of spotted hyenas (*Crocuta crocuta*) sit at a respectable distance of some thirty yards. They have been waiting there, patiently eyeing the zebra carcass, for several hours.*

Hyenas are much smaller than lions—about half the height and one-third to about one-half of the weight. It was once thought that these dog-like carnivores were mere scavengers, but although they do prefer scavenging to hunting, they are formidable predators in their own right. They are gregarious, living in clans composed of as many as eighty or more individuals. The female of the species is dominant over the male and is noted for her aggression. The clan is led by the dominant female. The remarkable feature of the female hyena is that she does not have normal female genitalia. The clitoris is enlarged to look like a penis, and the vaginal labia are fused to form a structure like a scrotum, complete with fatty deposits emulating testes. As a consequence, the female has the external appearance of a male.

Luckily for the lions, this is a small group, only half a dozen individuals; otherwise the hyenas would have snatched the zebra from the lions

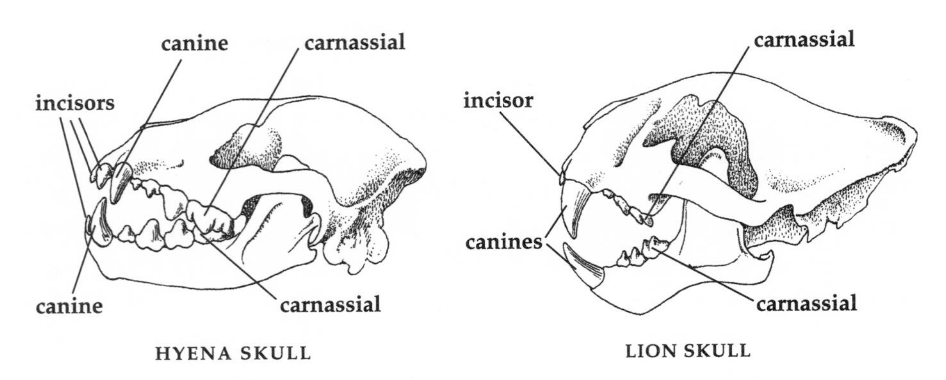

Hyenas with their massive, bone-crushing teeth, have more robust skulls than do lions.

while there was still meat on the bones. Lions will also snatch fresh kills from hyenas, and the outcome of the frequent battles between the two species is largely determined by which side has numerical superiority. However, the level of competition between spotted hyenas and lions is mediated by the fact that lions are primarily interested in meat, whereas hyenas will eat bones too. These differences in dietary preferences are reflected in their teeth.

Mammals, including ourselves, have three types of teeth: incisors, at the front, which are chisel shaped and often used for nipping; canines, flanking the incisors, which are typically dagger shaped and used for piercing and stabbing; and cheek teeth, at the sides, which may be flat or pointed, and which are used for grinding or for slicing. The cheek teeth are further categorized as premolars if they are replaced by permanent teeth, and molars if they are not represented by milk teeth. We have two incisors in each half of our upper and lower jaws, one canine (like all mammals), which is blunt and looks more like an incisor, and five cheek teeth (two premolars and three molars). Our cheek teeth are blunt, and the last one in each jaw half is called a wisdom tooth, because it does not usually appear until we are in our early twenties.

Most carnivores have specialized cheek teeth, called *carnassials,* with sharp cutting edges, for slicing through flesh. There is one carnassial in each half of the upper and lower jaws, at the back, arranged so that the top ones (last premolars) slice against the bottom ones (first molars). Carnivores have long, pointed canines, which are used for killing.

Hyenas can crunch through the biggest of bones as if they were pretzels, and the reason for this is immediately apparent when you look at their teeth and jaws. I have two skulls in front of me, one belonging to a hyena, the other to a lion. The skulls are about the same size (it is not a full-grown lion), but the hyena's is more heavily built, and most of its teeth are massive compared with those of the lion. Both have sharp carnassials, but whereas the lion has two cheek teeth in front of these, which diminish in size, the hyena has three large ones. The hind two of these are especially massive, with bluntly pointed crowns. These are the teeth the hyena uses for crushing bones. Its bone-crushing dentition is matched with an effective digestive system. Little goes to waste with hyenas, and it is reported that the people of Harar, in Ethiopia, encourage them to enter the town to consume the edible garbage, a role they appear to serve in other towns too.

*The male lion looks down at the zebra carcass, then across at the recumbent lionesses. Suddenly, out of nowhere, a bush duiker (*Sylvicapra grimmia*) appears, not five yards in front of him. Both look equally surprised at the presence of the other, but the lion's response is automatic. Closing with the small antelope in a single bound, the lion seizes it with both paws. Felid teeth sink deep into bovid neck; the lion is so big, and the antelope so small, that the head is almost severed. The lionesses, on their feet within moments, sniff at the fresh carcass, but none of the satiated lions make any attempt to eat it.*

Viewed through anthropomorphic eyes, the demise of the antelope is killing for the sake of killing. But there are sound biological reasons for the lion's action. Lions, like most other predators, are by no means successful every time they hunt, losing more potential prey than they ever catch. Nor can they always locate suitable prey when they want to, so they have to be opportunistic in their hunting. The difference between feeding or going hungry hinges on split-second timing, so their response to a potential quarry has to be instant. When lions are presented with a suitable target, the correct response has to be to attack. And when they have more meat than they can consume at one meal, as in this case, they will either stay with the kill for a few days, until it is eaten, or abandon it to the scavengers.

The male lion takes a last contemplative look at the two carcasses, turns slowly around, and walks away, followed by the rest of the pride. The hyenas, whooping and cackling and licking their lips, descend on the leftovers as if they were starving. The heaving, growling, shoving free-for-all lasts less than fifteen minutes, after which time all that remains of the two carcasses are bloodstains on the ground.

The lion might be the mightiest predator, but in many regards the wild dogs (*Lycaon pictus*), sometimes called hunting dogs, are more impressive. They are widespread across the African savanna but are fairly sparsely distributed, so they are not often seen by visitors to national parks. They are about the size of a large domestic dog, slightly built, with a lean body, long slender legs, and large rounded ears. Their coat is short, with white, fawn, and black patches. Like hyenas, they are gregarious, but differ in that only the dominant pair breed, and their offspring are raised communally in their close-knit society. Packs range in size from a few individuals to over forty, but average around ten. The pack usually hunts twice a day, generally between about six to eight in the morning and five to seven in the evening. Pups are raised in dens, and if too young to go hunting, they stay behind in the den with a parent, often with one or two other adults for additional security. Once all the youngsters are able to join the hunt, the pack no longer returns to the den at night, leaving its members free to roam far and wide.

A wild dog's lot is not a bad one. The dogs are rarely active for more than four or five hours a day, when they hunt and feed, the remainder of their time being spent at leisure. A typical day begins at first light. They stretch, yawn, and greet each other with much vocalizing and tail wagging. As time passes, the intensity of their socializing increases. There are lickings and nuzzlings and playful romps—all the good-natured behavior we associate with our own canine pets. It may be difficult to visualize these playful dogs as predators, but appearances can be misleading: they are the most formidable killers on the African plains, besides our own species.

This is a big pack. Over thirty dogs, working themselves up into a state of much excitement. A friendly fight breaks out between two young males—a romping, tugging, roll-'em-in-the-dust affair which spawns similar contests elsewhere in the pack. The revelry goes on for several more minutes, but then three or four of the older dogs, acting as leaders, strike off together at a trot, and the others are quick to fall in behind. The pack spreads out to advance on a broad front, a strategy to flush out any animals hiding in the tall grass. There is no attempt to keep in a line abreast; some dogs follow well behind the leaders, with a few stragglers bring-

The African wild dog is much smaller than the wildebeest.

ing up the rear. Maintaining a steady trot of 5 mph (8 km/h), they can cover a great deal of ground during a hunting foray. One of the stragglers stops to sniff at something in the grass and is joined by a companion. The first dog jumps high in the air, landing on all fours. When the maneuver is repeated, it has the desired effect. A small mouse scurries out of hiding and is immediately snatched up by the first dog and tossed across to the other. The mouse is unharmed, and the two dogs play with it for several minutes before one of them bites it and wolfs it down.

*The pack has covered about 2 miles (3.2 km) without spotting any suitable prey when it comes to a rise in the ground. In anticipation of surprising some unwary quarry on the other side, the dogs pick up the pace, bracing the hill at a brisk trot. There is nothing on the other side, but they do catch sight of a large herd of wildebeest (*Connochaetes taurinus*), a mile or so in the distance. This is of considerable interest to the dogs, and they change course to intercept the herd. Four hyenas that have been following the pack for the last mile also change course.*

An adult wild dog stands about 2 feet (0.6 m) tall and weighs about 50 pounds (25 kg), compared with a wildebeest's 4.5 feet (1.4 m) and 600 pounds (275 kg). This is a considerable size disparity, but the dogs have another problem to contend with, one having to do with their anatomy. Dogs, collectively called canids, are related to felids (cats), and while the two groups have much in common, there are some important differences between them. These differences are both anatomical and *physiological*, that is, pertaining to the internal workings of the body. As a consequence, canids have a radically different hunting strategy from that of felids. The lion is typical of other felids in that it stalks its prey to get as close as possible before making its charge. This is because the lion has a lower top speed than its prey but enjoys a superior acceleration. The charge is usually over a short distance, between one and two hundred yards (or meters), and ideally less. If the lion cannot capture the prey within that distance, it has to give up the chase because felids lack stamina. Canids, on the other hand, appear to have no shortage of stamina, as anyone who has a dog will know. (I'm pretty fit, but I cannot tire out my friend's young golden retriever.) Felids capture their prey by grasping it or swatting it with their front paws, usually toppling the prey to the ground. This requires both forelimb mobility, to enable the paws to be turned inward, and also a reasonable flexibility at the shoulder joint, so that the limbs can be moved from side to side. The limbs also have to be well muscled, for grappling with the prey, which makes them relatively heavier than those of canids.

An animal's body, like a piece of manufactured equipment, cannot be optimally engineered for all roles, and specializations in one area result in compromises elsewhere. Just as an aircraft cannot be designed to achieve high speeds and carry large payloads, so a felid's body cannot have evolved for optimal running performance and optimal prey capture. Its body design is therefore a compromise between the two competing sets of requirements. Felids have a good acceleration, partly attributable to their well-muscled legs. They have a modest top speed, which is not as fast as their prey, and impressive prey-handling capabilities. Canids, on the other hand, are more optimally designed for running, but at the expense of prey handling. Forelimb mobility is limited to the back-and-forth movements associated with running, and their legs have little lateral mobility, so most dogs cannot turn their front paws inward as felids can. A domestic cat can catch a mouse in its paws or swat at a ball of wool, but a dog's manual dexterity does not go much beyond a stiff-pawed handshake.

As a consequence of these anatomical differences, hunting dogs, and other canids, are unable to grasp their prey. Nor can they swat it, having to rely entirely upon their jaws for seizing. Due to these limitations, they must hunt in packs if they want to tackle animals larger than themselves. Solitary felids, in contrast, can catch prey much bigger than themselves, by virtue of their opposable paws. On the other hand, canids, with their lighter legs, can conserve more energy, making them more efficient runners. They also have far more stamina than felids. Wild dogs can keep up a chase for 2–3 miles (3–5 km) if need be, and their hunting forays can extend over distances of up to 25 miles (40 km). Felids, on the other hand, are exhausted after a chase of a few hundred yards, and, with the exception of the cheetah (*Acinonyx jubatus*), they avoid extensive wanderings. Indeed, canids usually have more stamina than their prey, enabling them to run them to exhaustion. Predators like felids, restricted by their stamina to short dashes of a few hundred yards, are described as *ambush predators* or *stalkers*, while those that can sustain a long chase are called *pursuit predators* or *coursers*. Hyenas, which are more closely related to felids than to canids, are pursuit predators, like canids. The biggest species, the spotted hyena, which is also the most common, is a powerful and most formidable predator.

Stamina involves supplying the skeletal muscles with sufficient oxygen to maintain high activity levels, and one of the requirements is a large blood flow from the heart. Predictably, dogs have bigger hearts than cats of similar size; they also have bigger lungs. Incidentally, heart size can be increased by appropriate training. This explains why joggers have larger hearts than those who do not exercise, and why marathon runners have bigger hearts than sprinters.

The wildebeest herd is over one hundred strong, giving the individuals a strong sense of safety. The bearded grazers see the approaching pack of wild dogs but are not unduly concerned because they are still one-quarter of a mile away. They carry on with their chomping and incessant grunting, without paying too much heed to the pending danger. At 200 yards (185m) the dogs reduce their speed to a slow trot, but the wildebeests still show little sign of interest, far less of concern. At something over 100 yards, the dogs come to a ragged halt and survey the scene. Several moments pass. The wildebeests closest to the pack stop feeding and stare with dumb-beast eyes at the diminutive dogs. The dogs stare back, a knowing twinkle in their eyes. More wildebeests stop grazing. Mild concern ripples through the herd. The panting dogs, wet tongues lolling, look from one to another, like mischievous boys plotting a prank. They lower their heads and walk slowly toward the press of wildebeests. Their head-lowered stalking posture, ears flattened, eyes fixed straight ahead, instantly transforms canid clown to ominous killer.

The wild dogs' strategy is to weed out a calf or sick adult, and to do this they must stampede the herd. Most of the wildebeests stop feeding, and loud snorts of alarm sound throughout the herd. But apart from some nervous hoof stamping and milling closer together, they stand their ground.

The pack, now less than 100 yards from the herd, breaks out into a fast trot. The herd stampedes. The wildebeests' defensive strategy is to pack close together as they run, keeping calves, pregnant females, and other weaker members toward the center of the herd, though the weaker ones inevitably fall back toward the rear. The dogs, for their part, have to break up the herd to get to these potential prey. Some of the dogs chase behind the herd, others run alongside, at speeds of about 25 mph (40 km/h)—the Apaches attacking the wagon train. Still others, risking death or serious injury from the flaying hooves, dart into the fray in their attempts to break up the formation. The chase goes on for over a mile, but the herd still maintains its close-packed integrity.

The dogs thrust and parry with no signs of fatigue, but with no sign of a suitable quarry either. Perhaps there are no calves or weaklings. If that is the case, the dogs will break off the attack and try again elsewhere. But they are not ready to give up the chase just yet. The herd wheels around to the right, and some of the dogs manage to cut around and charge toward the leading animals. This has the desired effect, and the herd is split into several divisions. The dogs press home their advantage and soon find what they have been looking for: a fairly large calf, accompanied by its mother, begins to drop back from the rest of the herd. Two dogs converge on the calf, causing it to veer away from the other wildebeests, still accompanied by its mother. The two wildebeests are still galloping at a fair speed, but the calf is beginning to tire. Three other dogs fall in behind the two pursuers. The lead dog closes with the calf and launches itself at one of its hind legs. Its sharp teeth make contact, and the jaws lock tight, bringing the calf to a faltering stop. A second dog grabs it by the nose. Two more tear at its underbelly, ripping it open and pulling out its gut. Its helpless mother looks on. She is fifteen times heavier than the largest dog and armed with a pair of sharp curved horns, like a cow's—horns that can kill. But although adults often do attempt to defend their young, this one will not.

Many of the *ungulates* (hoofed mammals) on the African plains have horns. Some of them are especially long and dangerous like those of the eland (*Taurotragus oryx*) and oryx (*Oryx gazella*), but they are not often used to defend themselves against predators. This is all the more remarkable considering that they are used in fighting among themselves. Why? The key to the paradox lies in understanding what horns are primarily used for, and this all has to do with natural selection and ungulate social systems.

An animal's primary goal in life, its very raison d'être, is to produce as many offspring as it can, passing its genes on to future generations. This is an animal's way of achieving immortality—not that it is a conscious act but rather a programmed evolutionary response. An animal's capacity to produce offspring is described as *fitness.* This evolutionary term should not be confused with the usual meaning of the word. In evolutionary terms, an animal that produces the most offspring is the fittest, regardless of its physical condition.

Many animals are polygamous, including most of the horned ungulates. Males compete with other males to sire as many females as possible, to perpetuate their own genes. The fierce competition for females is a strong selection pressure and has resulted in the evolution of an array of structures of adornment. These range from the peacock's flamboyant tail feathers, used to attract females, to the stag's imperious antlers, used in ritualized fighting with other males. In modern human society, such biological artifices as the male's manly physique are being eclipsed by material signs of his ability to provide for his mate and offspring—from expensive suits to Porsches.

The horns of a male antelope or mountain goat are not just decorative but are used in elaborate horn-locking and head-butting contests with rival males. In many species it is only the males that have horns, but in others (where males and females live together in herds yearlong) the females have them too, though these are often smaller than the males'. Horns, then, have evolved for use in aggressive interactions with other males. But why should they not also serve a defensive role against predators? Sometimes they do; lions, for example, are sometimes gored by buffalo. But the fact remains that ungulates, no matter how well armed they might be, would generally rather run than fight. And the reason they prefer to run is that they are so good at it! The point has already been made that most prey species can run faster than most predators, and to see why this is so requires a comparison of their skeletons.

A good indicator of an animal's top running speed can be obtained by comparing the length of its lower leg segment—from the elbow or knee down to the foot—with the length of the upper segment. In animals adapted for running, described as being *cursorial,* the lower segment is longer than the upper one, the disparity between the two being a good guide to relative speed. The zebra and the lion are both cursorial, but the elephant (*Loxodonta africana*) and hippopotamus (*Hippopotamus amphibius*) are not, and this fact is immediately obvious from a comparison of their

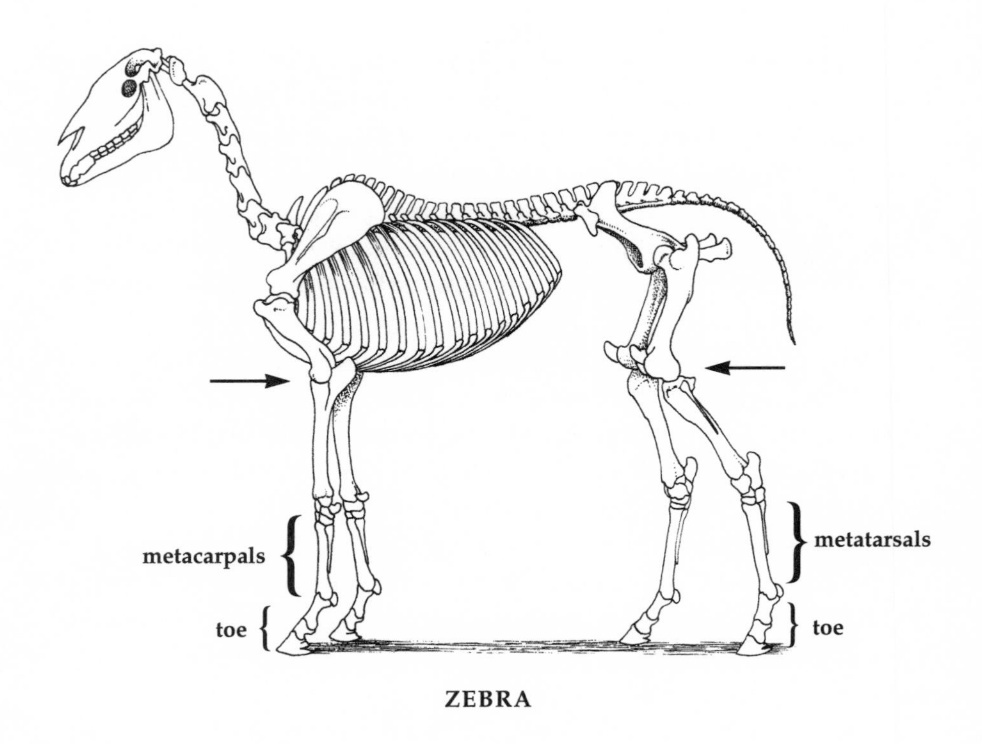
metacarpals
toe
metatarsals
toe

ZEBRA

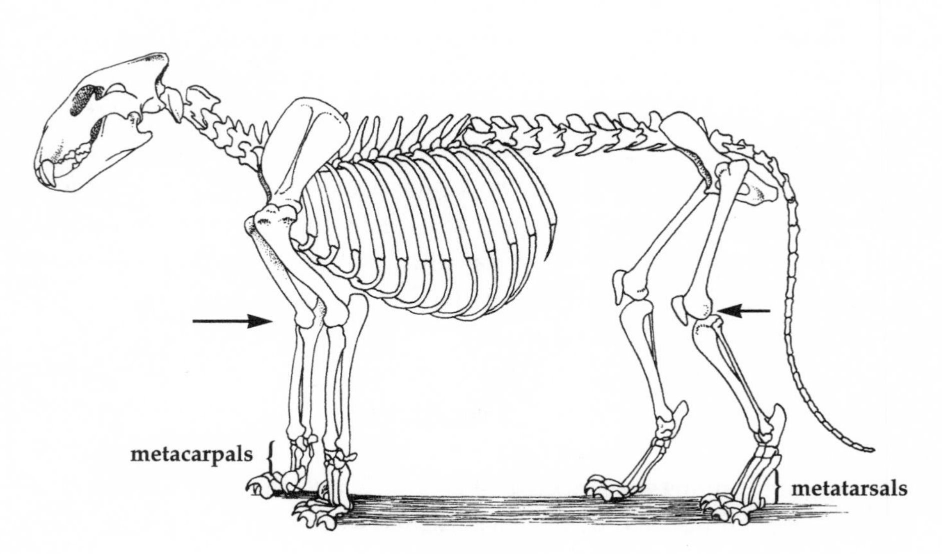
metacarpals
metatarsals

LION

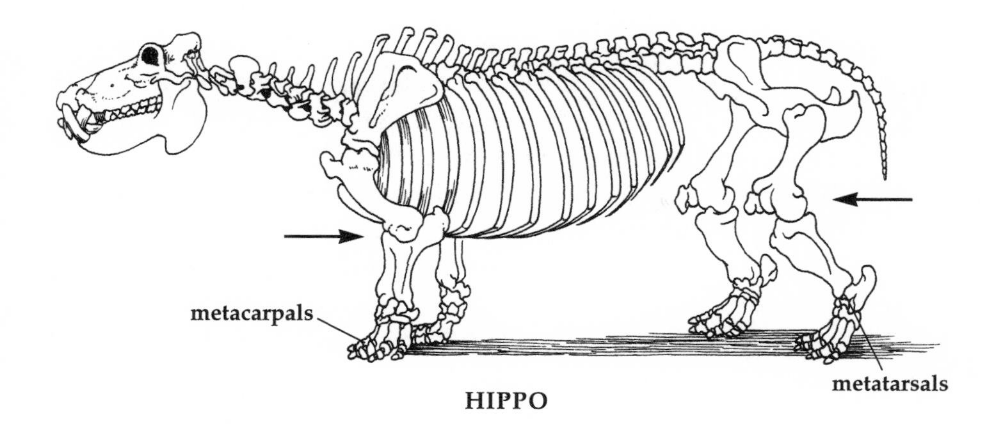

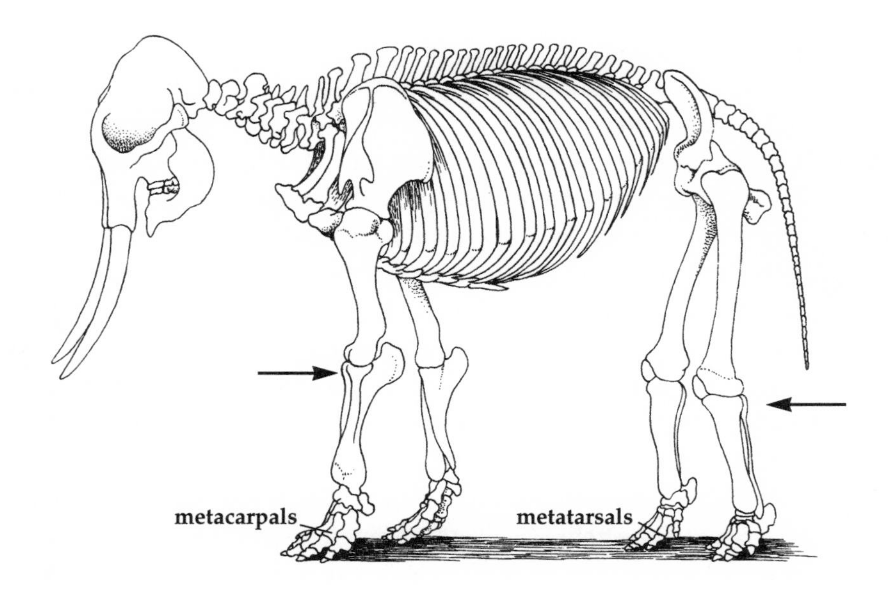

Some idea of the running ability of an animal is given by the length of the lower-leg segment (below the elbow and knee, shown by arrows) relative to the upper one. (The Indian elephant, Elephas indicus, *is depicted.)*

skeletons. The lower limb segments in the zebra and lion are almost twice the length of the upper segments, whereas they are more nearly equal in length for the elephant and hippo. Lengthening of the lower segment in the zebra has been achieved by an elongation of the foot region, from the wrist and ankle down to the toes. In the lion the lengthening is primarily in the shin and forearm, though the hind foot is fairly elongate too.

When we walk, our entire foot, from heel to toe, is placed on the ground. This style of walking is termed *plantigrade.* Dogs and cats, in contrast, walk on their toes, with the heel and palm raised off the ground. Their way of walking, called *digitigrade* for obvious reasons, effectively adds another segment to the lower leg. This segment is made up of the bones in the palm of the hand (metacarpals) and the sole of the foot (metatarsals). If this segment of the foot is long, as in cats and dogs and lions, it increases running speed, and for two main reasons. First, it increases the length of the *stride*—the distance between footfalls of the same foot—so that a larger distance is covered with each leg movement. Second, it increases the speed of the stride. This is because the speed of the foot is the sum of the speeds of the individual limb segments.

If you look at the illustrations on pages 24–25 you will see that the elephant and hippo walk on their toes, but their bones (metacarpals and metatarsals), unlike those of the lion and the zebra, are very short. Here walking on the toes has to do with weight support rather than with speed, because the elephant and the hippo have an extensive resilient pad behind their foot bones to cushion their heavy footfalls. Their posture is not truly digitigrade and is referred to as *subungulate.*

Cursorial adaptions have gone even further in the zebra than in the lion, because instead of walking on the balls of its toes, it walks on the toe tips, like a ballerina on pointe. And instead of having claws at the end of its toes, the zebra has hooves. In earlier times horses had three toes, but the side ones were lost several million years ago and the modern horse, as well as its close relative the zebra, walks on a single toe. Antelopes and cattle have two toes, hence two hooves. Walking on tiptoe is called being *unguligrade,* which is why hoofed mammals are called ungulates. Here the disparity between upper and lower leg segment is even greater than in digitigrade cursors like lions and wild dogs. This helps explain why most ungulates can run faster than most predators, the notable exception being the cheetah. [2]

Faster runners have other features in common besides longer lower leg segments. Their limb bones are more slender, and their legs are longer relative to the rest of the body. If you compare the skeletons of the elephant and hippo, you will notice that hippos have much shorter legs relative to their body. Which is the faster runner? Elephants, not noted for being fleet-footed, have a top speed of only about 22 mph (35 km/h), but hippos cannot go much faster than about 16 mph (25 km/h).

Felids and canids are both cursorial, but, as we have seen, the felids are adapted for short sprints, where acceleration is at a premium. Canids, on the other hand, are specialized for long-distance chases, where their superior running efficiency contributes to their endurance. And just as human sprinters are built like weight lifters while marathon runners are of slender build, so lions have heavy, well-muscled limbs while those of wild dogs are slender. Heavily muscled legs are good for acceleration, but the extra weight of all that muscle, together with the more robust bones to which they attach, adds considerably to the inertia of their limbs. *Inertia* is the tendency for an object to resist changes in its state of rest or motion. A light baseball bat, for example, has less inertia than a heavy one, so it can be swung faster. And if the two bats are to be swung at the same speed, the lighter one requires less energy. Limbs with high inertias are therefore more demanding of energy, which is one of the reasons that felids cannot keep up their high-speed chases for very long.

Let's return to the paradox of why well-armed ungulates tend to flee rather than fight. The first point to remember is that most ungulates are specialized for high-speed endurance running, whereas their predators are specialized for killing. The ungulates' logical defense strategy is therefore flight, though there are some exceptions, like the buffalo and the hippo, which are both formidable adversaries. The effectiveness of the flight strategy is reflected in the fact that the odds of being caught are usually well in favor of the prey, the hunting successes for predators generally being less than 50 percent. [3] Since feline predators are reliant on their superior acceleration, they are only dangerous at close quarters. So if an ungulate is vigilant and keeps a safe distance from prowling felids, it is not at risk. The threat from canids and hyenids (hyenas) is somewhat different because they have stamina and can run any animal down—like the goblins in *The Lord of the Rings*. And when they catch up with their prey, the outcome is inevitable, so there is little point in an ungulate attempting to retaliate. This, together with their state of exhaustion, probably accounts for the passive resignation with which most animals face death at the hands of wild dogs and hyenas. Attacks from felids, on the other hand, usually happen so quickly that the prey seldom has a chance to defend itself, even if it wanted to. There are exceptions, of course, and some animals do successfully fend off predators, but not very often.

George Schaller, in his wonderful book *The Serengeti Lion,* provides an excellent example of an ungulate, namely a buffalo, capitulating to predators. The African buffalo (*Syncerus caffer*) is a big and dangerous horned animal that will attack lions. It is therefore frequently avoided by lone hunters, but some lions, hunting in packs, actively seek out this species. Schaller reports seeing a bull buffalo, badly wounded from a previous lion attack, standing in mud and water up to its belly. It is surrounded by a large pride of fourteen lions, but they are reluctant to get wet by going in the water after it. An hour and a half passes, then five nomadic (not belonging to a pride) males arrive, driving the other lions away. They also surround the buffalo, without venturing into the water. Fifteen minutes pass, then the buffalo leaves the water, on his own volition, and walks slowly toward the lions. Schaller describes the act as a "suicidal gesture," and I think he is probably right.

What other defensive strategies do prey species have, besides fleeing? One of their most important attributes is vigilance. When predators are in the vicinity, it is important to know this so that appropriate measures can be taken. Sight is obviously an important sense, and prey species typically have laterally placed eyes. The visual field of the vertebrate eye is about 180°, so placing the eyes on the sides of the head affords good all-around vision. This is enhanced by the bulbous shape of the front of the eye, protruding from the head like a fish-eye lens. Hence it is almost impossible to sneak up behind a horse without being seen. Tall ungulates, like zebra, eland, and buffalo, have an unobstructed view above the grass, whereas short animals, like the smaller antelopes, often do not. Grant's gazelles (*Gazella granti*), which are both taller and more alert than Thomson's gazelles, detect approaching predators from greater distances. Consequently, when the two species are grazing together in mixed herds, Grant's gazelles provide an early warning of danger to their smaller relatives. Hearing is probably also important, and an ungulate's long mobile ears facilitate directional capabilities, allowing them to detect the direction of the sound. The sense of smell may be of less importance in detecting predators, judging from the negligible effect that wind direction has in the hunting success of lions. However, cheetahs nearly always hunt downwind of their prey, so olfaction is obviously important here. Cryptic coloring also plays an important role, especially in those species that live in forests and that do not move around in herds.

Once a predator has been detected, the appropriate response is mediated by the identities of the antagonists. Just as predators size up potential prey to decide whether, and how, to attack, so the recipients of their attentions make their own judgments on how to respond. One immediate response is to make any necessary adjustments in the safety distance between themselves and the predator. A herd of zebras might allow a cheetah to approach within 65 feet (20 m) without fleeing because cheetahs attack them only occasionally, but they double the distance for a lion. If young calves are present, the herd might flee before the predators approach to within 110 yards (100 m). The presence of offspring also affects the behavior of parents. A female Thomson's gazelle with a calf is likely to be emboldened to attack a small predator, like a jackal, by butting it. But her response to a hyena would likely be limited to her trying to distract its attention from her calf.

Prey species must be vigilant at all times to avoid placing themselves in vulnerable situations. Thickets and other potential hiding places for predators are usually approached with caution, and watering holes, which are favorite ambush sites, are avoided at night. If an animal is unfortunate enough to be pursued by a predator, it can often escape by adopting an erratic, zigzagging flight. Some antelopes leap high into the air as they run, the impala (*Aepyceros melampus*) being one of the most adept performers, with spectacular leaps as high as 10 feet (3 m). These athletic animals can dodge and change directions with lightning speed, sometimes even turning around and leaping right over the heads of their pursuers. Many antelopes have a defensive strategy called *stotting* or *pronking*, a distinctive bounding gait where they launch themselves, stiff legged, from the ground. Stotting usually begins at the first sign of danger and is generally used against pursuit predators like hyenas and wild dogs, rather than against ambush predators like lions and cheetahs. [4] The sight of a herd of antelopes bouncing about in all directions must cause considerable confusion to an onlooking predator, and wild dogs apparently do not chase stotting animals.

Many ungulates are gregarious, and one of the obvious benefits of living in a herd is increased awareness. An animal cannot spend its entire day on the lookout for predators; otherwise it would starve to death. However, at any given moment there is likely to be at least one vigilant member in the herd, and any concerns this individual might have are immediately communicated to the rest of the herd, usually by vocalization. Herding can also be used to protect calves and other vulnerable members, as in the packing behavior of the wildebeest. There might also be some value in the confusion effect that a large number of moving animals can have on a predator. You can experience this effect for yourself by joining in a game of tag with a number of children (or adults, but I've always found them less accommodating to my eccentricities!). If you try chasing one particular individual who is in the middle of a knot of others, you find yourself changing your mind as other targets come closer to hand. But as soon as somebody breaks away from the crowd, he or she becomes a much easier target. This may explain why it is often the ungulate that acts differently from the rest of the herd that gets chased. Just looking different from the others may attract a predator's attention, and this was demonstrated when a researcher painted a wildebeest's horns white; the marked individual soon disappeared from the herd. Predators can probably be very single-minded, though, locking onto one target regardless of any other opportunities that may arise during the chase. Although herding behavior has some obvious benefits, there is one disadvantage, especially for large herds: they can be seen from a long way off, attracting all manner of predators to come and feast. Solitary ungulates are less visible, and some, like the reedbucks (*Redunca arundinum*), have a defensive strategy of hiding in tall grass. There are therefore two broad categories of defense: either ungulates are social and visible, using the collective vigilance of the herd to warn them of danger, or they are solitary and cryptic, using camouflage and concealment to keep them from harm.

Over the course of time—many millions of years—predators have evolved more effective ways of capturing and killing their prey. Their potential quarry, for their part, have also undergone change, evolving more effective ways to avoid being caught. This has led to the idea that the protagonists are linked together in an arms race, improvements in offense being countered by defensive countermeasures. This is an entirely plausible notion, but is there any evidence that this has happened?

The fossil record for mammals during the last 50 million years is quite impressive, and it is possible to trace the evolutionary trends that have occurred among the ungulates and their potential predators. One of the trends was for ungulates to evolve longer legs, becoming more fully cursorial. Indeed, ungulates with leg proportions similar to modern cursorial species appeared about 20 million years ago. But what about the contemporaneous predators? Paleontologist Robert Bakker, investigating the problem in 1983, came to the conclusion that the predators always lagged far behind the ungulates in their cursorial adaptions. Since the predators were not evolving fast enough to track the changes taking place in their prey, the arms-race analogy is not supported. A decade later, Christine Janis, a paleontologist who specializes in the study of animals of this time period (called the Cenozoic), and Patricia Wilhelm took an in-depth look at the problem. They concluded that pursuit predators comparable to the present-day wild dog and wolf did not appear until about 2 million years ago. There were plenty of other predators around during those far-off times, including a wide variety of sabertooth cats, but they were mostly ambush predators, like the lion. And none of them were as specialized in their cursorial adaptions as modern species, being more akin to a cross between a lion and a bear in this regard. Like bears, they were probably powerful, but they were not well adapted for running. The absence of fleet-footed predators raises the question of why the ungulates of the time were so far advanced in their running abilities. Why be a gold medalist if the competition cannot even qualify for the race? The probable answer has to do with ungulate diets, and with the climate changes that were taking place at that time.

Leaves and grass, in contrast to meat, are not very nutritious; nor are they easy to digest. A good rib roast, or dead impala depending on your tastes, is digested in a matter of hours, broken down by enzymes produced by the body. A vegetarian meal, in contrast, takes considerably longer to digest. The bulk of plant material—fiber and the substance forming the cell walls—is made up of cellulose, which cannot be digested by any enzymes produced by vertebrates. The only way it can be handled is to leave it in the gut for several days while it is fermented by bacteria. A similar process takes place inside a garden composter, but that takes considerably longer. Since plants are so low in nutrition, they have to be eaten in large quantities, and that requires a large gut. Herbivores therefore have much larger guts than carnivores, which explains why cows and horses are so broad compared with dogs and lions. Ungulates have evolved complex multichambered guts for the fermentation process, thereby improving the efficiency of the extraction of nutrients. These evolutionary changes were probably taking place during the early part of the Cenozoic Era, a geological time period that began 65 million years ago. The Cenozoic is often referred to as the Age of Mammals because they were the dominant land vertebrates. It followed immediately after the demise of the dinosaurs and the Age of Reptiles (Mesozoic Era). The climate at the beginning of the Cenozoic was tropical to subtropical at higher latitudes, and much of the land was covered by lush vegetation with dense forests. As time progressed, the seasons became more different from one another, and the higher and the temperate latitudes experienced winter frosts; the second half of the Cenozoic saw periods of summer droughts. Grasslands spread across much of Africa, Eurasia, and North and South America, at the expense of the forests. These environmental changes were accompanied by the evolution of a diverse array of grazers—animals that feed mainly on grass—which exploited the vast seas of grass that covered so much of the land. The earliest ungulates were browsers, feeding on plant material other than grass and living primarily in the forests. The browser's diet of leaves, shoots, berries, and twigs is generally more nutritious than grass, and since this food source grows on trees and bushes rather than on the ground, it is less thinly spread out. Consequently, browsers do not have to travel as far as grazers to gather their food, though it must be pointed out that many grazers, like the hartebeest and Grant's gazelle, do not travel far either.

Drawing these facts together—the spread of grasslands, the exploitation of the new resources by grazers, and the grazer's need to travel far—Janis and Wilhelm suggested why ungulates became more cursorial: it was simply to reduce their transport costs. Wildebeests, for example, may travel thousands of miles across Africa in search of fresh pastures, and the amount of food required to fuel these wanderings is substantial. Any improvement in their locomotor efficiency is therefore at a premium, which is precisely what is achieved by becoming more cursorial. We can never know for sure what happened during the remote past, but the idea that the evolution of cursoriality among ungulates was related to increasing locomotor efficiency, to reduce transport costs rather than to escape from predators, makes a great deal of sense to me. The fact that being cursorial facilitated escape from the fleet-footed predators that evolved later does not detract from the idea.

The fossil evidence clearly does not support the idea of an arms race between predator and prey, but there are some logical arguments against the notion too. For example, suppose a prey species evolved some improvement in its antipredatory arsenal—better acceleration, improved eyesight, or whatever—that reduced its chances of being captured. Initially the predators would catch fewer prey, and the prey population would accordingly increase in size. But as they became more plentiful, this would work in the predators' favor; the prey are still hard to catch, but there are more of them to chase. The cost-benefit situation also has to be considered. Nothing in life is free, and any improvements in one aspect of an animal's performance incurs costs elsewhere. Suppose a prey species improved its acceleration by increasing the size of its leg muscles. The improvements in its escape performance would decrease its chances of being caught, but the additional weight of the extra muscle mass would increase its transport costs, causing the prey to consume more food. The same argument applies to the predator too. On the other side of the coin, let us suppose the predator improved its hunting success but there was no reciprocal response from the prey. How detrimental would this be to the prey species? This raises the question of what impact predators have on prey populations.

The interactions between predators and prey are complex and have many variables, all tempered by local conditions, as the following example will illustrate. Dale Seip, a Canadian biologist with the Ministry of Forests, made a study of wolf (*Canis lupus*) predation on moose (*Alces alces*) and caribou (*Rangifer tarandus*) in two remote areas of British Columbia: Quesnel Lake and Wells Gray Park. Throughout the year, the wolves and moose shared similar habitats in both areas, but the caribou spent their winter months in the highland, thereby avoiding predation by the wolves. However, the behavior of the caribou differed between the two areas during the summer: the Wells Gray Park caribou migrated to the mountains, while the Quesnel Lake population mingled with the moose and the wolves. As a consequence, the Quesnel Lake caribou population is declining, by about 25 percent of the population per year, while the other population is slowly increasing. The impact of predators is obviously highly variable and heavily influenced by local circumstances.

One of the factors influencing a predator's effect on prey populations is the number of sick and immature individuals that are taken. Animals die of disease and injury as well as by predation. Mortality rates are highest in young individuals, which is why females produce so many offspring during their lifetime. If a predator kills a sick or an immature individual, the negative impact on the population is minimal because that particular individual had less chance of breeding and contributing to future generations anyway. The extent to which this factor plays a role varies among predators and among prey. Schaller found that lions, for example, kill zebras and wildebeests of all ages, whereas cheetahs, leopards (*Panthera pardus*), and wild dogs concentrate on young individuals. And while lions and cheetahs primarily kill healthy animals, hyenas tend to focus on sick ones.

Hunting success varies among predators and the prey they hunt and is influenced by several factors, including the number of predators, the time of day, and the environmental setting. Schaller provides a wealth of hunting statistics for the Serengeti region, and a few examples will illustrate the point. The hunting successes of the lion, averaged for all hunts both day and night, varied from 14 percent for reedbuck to 32 percent for wildebeest, so the odds are well in favor of the prey. Hunting dogs and cheetahs did considerably better, with average successes for all prey hunted of 70 percent. When two lionesses hunted together, their success rate was approximately double that of a lone hunter, but no significant improvement occurred when more than two lions participated. However, a study of lions in Namibia showed that hunting success increased with the number of participants. The study site was in a flat, semiarid area, with very little ground cover, unlike the Serengeti, making it difficult for the lionesses to sneak up on their prey. The most successful prides were those in which the attacks were coordinated: Some of the lionesses encircled the prey while others lay in ambush. The ambush would not be sprung until all of the predators were in place, and each lioness specialized in her particular role. This level of planning and coordination is the work of an intelligent being. As with lions elsewhere, nocturnal hunts were more successful than those conducted during daylight, with successes highest on moonless nights.

Although lions, on average, are far less successful than wild dogs and cheetahs, they take a far greater share of the biomass (total weight) of prey. This is largely because lions are far more common than dogs or cheetahs. The impact that lions and other large predators have on the populations of prey species can be quite remarkable, as was shown by a study in Kruger National Park, South Africa. The most heavily predated species was the wildebeest, whose losses were estimated at a staggering 42 percent of its total biomass per annum. Impala, much sought after by cheetahs, lost 16 percent, while zebra losses were calculated at 14 percent. Predictably, the largest prey species, the buffalo, giraffe (*Giraffa camelopardalis*), and hippopotamus, were least heavily predated, their losses totaling just 5 percent. Lions accounted for just over half of all the biomass of all species killed. The carcasses left behind by the lions satisfied much of the food needs of hyenas. The Kruger hyenas, unlike those of the Serengeti, never robbed lions of their kill. Instead, they waited patiently for the lions to finish, sometimes for several hours, before moving in to devour the leftovers. This difference in behavior had less to do with good upbringing than with the fact that the hyena groups were much smaller than the prides. While demurring to lions, the hyenas did rob cheetahs of their prey, cheetahs being more lightly built and less aggressive than hyenas. Hyenas will also rob

The cheetah is the fastest land animal.

leopards given the chance, leopards being of similar size to cheetahs. However, leopards are excellent climbers and haul their kills into trees, where hyenas are unable to go.

Cheetahs, the fastest of all land animals, can reach speeds of about 60 mph (96 km/h), with reported speeds as high as 70 mph (110 km/h). Like other ambush predators, they use cover to stalk their prey, but because of their remarkable speed, they do not need to get so close. In the absence of cover, they will approach their prey in full view, suddenly accelerating to full speed. They hunt alone or in small groups, but are unable to defend their kills against large aggressors like lions and hyenas. This may explain why they sometimes hunt during the full heat of the midday sun, when most of the other large predators are resting. They are sorely persecuted by lions, which attack them and their cubs without provocation or apparent cause. Lions make a particular point of seeking out cheetah cubs and killing them, but not for food because the dead cubs are not eaten.

Leopards are nocturnal, hauling up in trees during the day, thereby drawing less attention to themselves. They are the ultimate ambush predators, stalking within yards of their prey and pouncing upon them before they have a chance to flee. They are the most widespread of all wild felids, occurring throughout much of Africa, the Middle East, and Asia. Where their territories overlap with human habitations, some leopards develop a taste for domestic pets. Their predilection for dogs is sometimes transferred to the dogs' owners, and prowling leopards have been known to slip through open windows, snatching people from their beds. Fortunately for our own species, this is a rare occurrence.

The predators we have been dealing with here are all related to one another and are classified as members of the same mammalian group, called the Carnivora. This large group contains a diverse array of animals including bears, pandas, badgers, mongooses, otters, and weasels. Most carnivores are predators, eating a wide range of other animals including mammals, reptiles, birds, fishes, and insects. But some, like the pandas, have become secondarily vegetarian. They retain the basic features of their predatory ancestry, including a short gut, claws, and long canine teeth, but these, like the rest of their teeth, are blunt, being specialized for grinding rather than for slicing. Plants, we have seen, take a long time to digest and require a long specialized gut for the process. But pandas, being carnivores, do not have a long gut, and as a consequence their digestive efficiency is low. In terms of their digestive efficiency, pandas are victims of their own inheritance—the carnivores' short gut—and provide a good example of how an animal's evolutionary past determines its present, and future. The study of the relationships among organisms, called *phylogeny,* is helping us to make better sense of our living world. In other chapters we will see many more examples of how an organism's phylogeny, its pedigree, acts as a constraint, limiting the options available.

Predators take much heavier tolls of prey in the northern latitudes of North America than in Africa because there are far fewer prey species. In some areas wolves, grizzly bears (*Ursus arctos*), and black bears (*Ursus americanus*) account for over 80 percent of all calf deaths among caribou, wapiti (*Cervus canadensis*), and moose, and are probably the main cause of adult deaths too. Bears are *omnivorous,* meaning they eat a variety of foods, both plant and animal. They are content to grub around for berries and roots and similar unappetizing things, pulling the occasional fish from the lake, but they won't pass up an opportunity to take an innocent warm-blooded prey either. Bears are ambush predators, expert at stalking their prey through the woods, and in spite of their ungainly loping gait, they have a remarkable turn of speed when they charge. Lacking the felid killing bite, they maul their prey to death, though some bears do use a paralyzing bite to the face. A biologist in Norway reported this killing technique being used on sheep by one particular bear. The bear seized the sheep by the nasal region, puncturing bone with its canine teeth and thereby inducing paralysis, presumably by the sudden damage to the sensory nerve trunks in the head. Once the sheep were paralyzed, the bear set to work on them, but there was one documented case where the individual recovered from its bite when the bear left it to attend to another sheep.

Sabertooth cats, mentioned earlier, were quite unlike any predators we know today, both in appearance and probably also in the way they hunted. They are characterized by the presence of remarkably long upper canine teeth, projecting down from the sides of the mouth like a walrus's tusks. There were many different kinds of sabertooth cats, belonging to several distinct and distantly related groups, and they appeared at different times during the last 30 million years. One of the best-known examples is *Smilodon,* a true cat, which lived in North and South America during the last 2 million years, becoming extinct at the end of the Ice Age. It was about the size of a lion but was more robustly built, giving the impression of great strength. The long upper canines are gently curved, and flattened from side to side. The jaw joint was modified so that the lower jaw could be swung wide open, giving maximum clearance for the sabers to do their work. Just how the sabertooths used their sabers has been debated since the time of their discovery. Paleontologist Blaire Van Valkenburgh pointed out that, being so long and thin, the sabers would have been vulnerable to breaking, especially if they became wedged in bone. It seems most likely that sabertooths ambushed large prey. There were plenty of large land mammals about at the time, like the giant ground sloths, and the sabertooths could have killed them by inflicting deep slashing wounds to the soft parts of the body, where the cats' sabers would not have contacted bone. The neck vertebrae of *Smilodon* are tall and robust, indicative of powerful neck muscles used for driving the head with sufficient force for the sabers to puncture the hide and penetrate the body. Their strategy may have been to inflict deep wounds, retreating to avoid injury, and waiting for their prey to bleed to death. One possible scenario, suggested by the discovery of the remains of baby mammoths in what appeared to be a sabertooth den site, was that the cats attacked young mammoths. By choosing their moment carefully, they might have inflicted a deep wound on the offspring, then retreated before the mother had a chance to drive them off. The sabertooth could then bide its time while the baby bled to death, then return to the carcass after it had been abandoned by its parent.

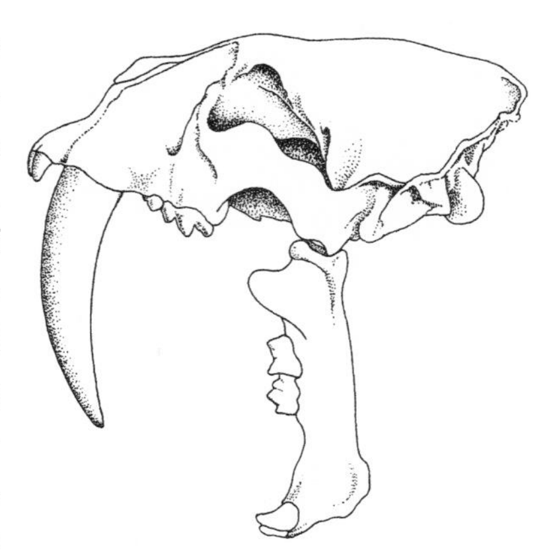

The skull of Smilodon.

From warm-blooded killers of the past we will now turn to some cold-blooded killers of the present.

2

In Cold Blood

Museums are odd places. I think I knew that before I took up my present position, but my first few days at Canada's premier museum, the Royal Ontario Museum, were quite unforgettable. I arrived from England at the height of the summer, to temperatures and humidity levels I would have expected in the tropics, not in Toronto. Many of the curators were out in the field, including my new boss, so I found myself in the unusual situation of being the most senior person in the department but having no idea of what went on there.

I expected to see the dinosaurs, and all the other very old dead things that a department of vertebrate paleontology was supposed to have, but I did not expect to see the *live* snakes and the live lizards. They were part of a new display in the dinosaur gallery, I was told, to give our visitors an idea of what real-live reptiles look like. And since it was a dinosaur gallery, the live reptiles had to be big ones. The lizards—monitors with particularly mean dispositions—were over four feet long, and the python was eight feet.

Being summertime, Parliament was not in session, and the news media were suffering the usual dearth of hard news stories. You can therefore imagine the coverage the museum got when the python turned up missing one morning!

"What if it comes across a child in the gallery?" one reporter asked, no doubt imagining the discovery of a contented snake with a swollen belly.

"I hope the child doesn't hurt it," replied Frank Ross, who looked after the reptiles. "A child could do more harm to the snake than the snake could do to the child."

The reporter, unconvinced, still clung to the notion of the python stalking the public galleries in search of innocent victims. It might even get out of the museum and wreak havoc on the people of Toronto.

"It's probably coiled up asleep somewhere," Ross continued. "Out of everyone's way. It'll turn up."

The python did indeed turn up a few days later, coiled up in an air duct just a few yards from where it had disappeared. Public safety had never been at risk, but if the escapee had been a panther rather than a python, the situation would have been entirely different. That is because of the physiological differences between reptiles, which are cold-blooded, and mammals and birds, which are warm-blooded. It is important to take time to look into these fundamental differences.

All living cells in an animal's body require energy to power the various chemical processes going on inside them. This energy is ultimately supplied by the food that animals eat. These chemical processes are collectively referred to as *metabolism,* and one of the by-products of metabolism is heat. Metabolic rates vary throughout the day; they also vary between individuals, and between species. Our metabolic rate is at its lowest during the night, when we are asleep, and it rises with increasing activity levels. It is higher when we are walking to work than when we are sitting at our desk, and would be higher still if we went jogging. The release of energy inside the cells usually requires oxygen, so when metabolic rate increases, so does oxygen consumption. Metabolic rates could be measured by assessing the amount of heat an animal gives out, but it is far more convenient to determine the rate at which oxygen is being used.

You may have seen athletes on treadmills wearing face masks, to measure the amount of oxygen being used. Similar equipment has been modified to measure metabolic rates for various animals, ranging from turtles to elephants. Fitting a face mask to a small animal like a lizard or a mouse is generally impractical, so the animal is usually placed inside a chamber, and the air passing through the chamber is analyzed instead. Metabolic rates vary with an animal's activity levels, so they have to be measured under standardized conditions; otherwise the results would not be comparable with others. The subject is kept at rest, at a comfortable ambient temperature, and the resulting metabolic rate is the *standard,* or *basal, metabolic rate.* Warm-blooded animals (birds and mammals) have basal metabolic rates about five to ten times higher than those of similarly sized cold-blooded ones (reptiles, amphibians, and fishes). And it is precisely because birds and mammals have such high metabolic rates that they are able to keep their bodies warm.

The terms *warm-blooded* and *cold-blooded* are still in everyday use, but they are not very precise. Anyone who has handled a snake knows this because a snake's body actually feels quite warm. But very little of the snake's body heat originates internally, from its cells, most of it having been supplied from the outside, either by the sun or by a heat lamp. Instead of referring to reptiles as cold-blooded, they are best described as *ectothermic,* meaning "outside heat." Similarly, birds and mammals are said to be *endothermic,* meaning "inside heat." There are advantages and disadvantages to each thermal strategy. Reptiles are usually sluggish first thing in the morning, their body temperatures having dropped during the cool of the night. Accordingly, they have to bask in the sun to raise their body temperatures, but once they have warmed up sufficiently, they can go about their business. By alternating between the sun when they are too cool, and the shade when they are too warm, many reptiles are able to maintain their body temperatures at optimum levels of about 35°C (95°F) or more. Endotherms, on the other hand, maintain temperatures of about 37°C (98°F) all the time, [1] so they are always ready for action.

I used to keep a small caiman (a crocodile). He had very sharp teeth, and I had to be careful how I handled him during the daytime, when he was warm. But I could do whatever I wanted at night, when he was cold, without any fear of being bitten. The obvious disadvantage of being ectothermic is that the animal's activity levels are dependent upon the environment. But its low metabolic rates mean that it requires far less food, which is an advantage. I used to feed the caiman a tiny piece of liver once a week, whereas the family cat demanded three meals every day. We should therefore not think that reptiles are inferior to mammals and birds; they are just different. Reptiles living in temperate climates, for example, avoid the problems of trying to find food during the winter months simply by hibernating. Most mammals, in contrast, have to forage in the snow to find sufficient food to keep their endothermic fires burning.

Reptiles lack any body covering like fur or feathers (though most of them have scales), their skin is dry and waterproof, and most lay shelled eggs on the land. They were the dominant land vertebrates (animals with backbones) during Mesozoic times, but their numbers have dwindled—

though they still outnumber mammals—and they are represented today by four main groups: the lizards, snakes, crocodiles, and turtles. [2]

While reptiles are the focus of this chapter, mention will also be made of amphibians: frogs, salamanders, newts, and their allies. Unlike reptiles, amphibians do not have waterproof skin, so they live in damp places or in water, and most of them are cold to the touch because they tend to avoid the sun. Their eggs, lacking shells, are almost always laid in water.

Mammals have stamina, enabling them to chase their prey, sometimes over long distances. Most reptiles, in contrast, have such low stamina that they are unable to chase their prey over anything but the smallest of distances. But that does not mean they cannot move rapidly when they want to—just try catching a lizard in the heat of the day. There is an obvious distinction between short bursts of rapid activity that cannot be sustained (sprinting) and long periods of sustainable activity (long-distance running). The underlying differences are physiological and have to do with the way that energy is supplied to the skeletal muscles.

Imagine being on a treadmill, wearing a face mask connected to oxygen-measuring equipment. You are walking leisurely, at the slowest setting, and your oxygen consumption is fairly low. As the speed of the treadmill increases, your metabolism increases, and so your oxygen consumption goes up accordingly. Your muscle cells are getting all the oxygen they need to break down the carbohydrate fuel they are using, so their metabolism is said to be *aerobic* (with air). You are now jogging along at a comfortable speed, still supplying ample oxygen to your muscle cells, so your exercise is still aerobic. Your oxygen consumption will keep on rising as the speed increases, but a point is eventually reached when an increase in speed is no longer associated with a corresponding increase in oxygen consumption. That is because your heart and lungs have reached their maximum capacity for supplying oxygen to the muscles via the bloodstream. But you are still able to run a lot faster. Your speed keeps increasing and eventually you are running flat out, but your oxygen consumption has not changed. The reason your muscle cells can keep on firing without getting the extra oxygen they need is that they are able to break down carbohydrates in the absence of oxygen. This is called *anaerobic metabolism* (without air). Sprinters perform almost their entire 100-meter race anaerobically, whereas long-distance runners perform most of their event aerobically. Anaerobic metabolism is far less efficient than aerobic metabolism but is capable of generating very high quantities of energy for short periods of time. It also generates lactic acid as a by-product. This accumulates in the muscles and is inevitably associated with muscle fatigue and general exhaustion. That is why sprinting cannot be kept up for very long. After sprinting, the lactic acid that has accumulated in the muscles has to be broken down, and this requires oxygen, which explains why we puff and pant after a hard run. The lion's charge and the cheetah's sprint are largely anaerobic, whereas the long-distance pursuits of African wild dogs and wolves are mostly aerobic.

Because reptiles have much lower metabolic rates than endotherms, their aerobic performances are very modest; they cannot cover such long distances as mammals, nor can they move as fast aerobically. Predatory reptiles must therefore rely largely on anaerobic performance to capture their prey. They can be as fast as lightning, but they lack endurance. Hence most reptiles adopt a sit-and-wait strategy, stealth and patience being of the essence. But some of them have evolved the most formidable weapon systems for seeking and destroying the target. The clinical war dispensed by reptiles appears far more sinister than the hot carnage served up by mammals. But we will begin with a most unsophisticated killer, the Nile crocodile (*Crocodylus niloticus*).

Crocodiles occur throughout most of Africa, from Egypt to South Africa. They are the closest reptilian relatives of the dinosaurs and retain something of the appearance of a predator from the remote past. Stories of their reaching lengths of 30 feet (9 m) are grossly exaggerated, but lengths of 20 feet (6 m) and more were recorded at the beginning of the century, before overhunting decimated their populations. The largest crocodiles found in Africa today are about 15 feet (4.6 m) long, but a reptile of this size can weigh upward of 1,000 pounds (450 kg) and is a most formidable predator.

Crocodiles spend half of their time in the water, seldom straying far from the water's edge. Most of their days are spent basking in the sun or lying in the shade, and if air temperatures get too high at the height of the day, they slip into the water to cool off. By alternating between sun, shade, and water, crocodiles can maintain their body temperatures at about 25°C (77°F). This is about ten centigrade degrees lower than the daytime temperatures of most reptiles and is probably correlated with their aquatic way of life. When the sun goes down and air temperatures drop below those of the water, the crocodiles slip back into the river or lake, where they remain until dawn.

When seen sprawled motionless on a riverbank, or floating lifelessly in the water, it is difficult to believe these reptiles capable of any physical activity. But crocodiles can move with astonishing speed when they want to—fast enough to pluck a bird from the air. They are opportunistic hunters, like most predators, and their catholic tastes range from crabs and snails to fish and fowl. There is a well-marked trend in food preferences according to body size. Smaller individuals feed largely on invertebrates—mostly insects and crustaceans—together with small vertebrates like frogs and toads. Birds and fishes are caught by individuals of most sizes, whereas the largest crocodiles prefer eating reptiles and mammals. Their reptilian prey include monitor lizards, pythons, cobras, turtles—shells and all—and other crocodiles, while their mammalian menu includes zebras, wildebeests, waterbucks, hippos, baboons, and the occasional human. And it is in hunting large prey like zebras that crocodiles are at their most chillingly cold-blooded.

Lakes and rivers are unavoidable for many of the ungulates on the African plain, whether as a source of water or as an obstacle to be crossed during migration. All the crocodile has to do is bide its time and a suitable prey will eventually come along. If the opportunity arises, a crocodile will attack animals on land, but its natural element is the water, and its usual strategy is to lie partially submerged and wait.

A herd of wildebeests approaches a placid lake in the middle of a hot afternoon. Aware of the dangers of being ambushed, they approach the water's edge with some trepidation. Once they begin drinking the cool water, however, caution gives way to thirst. The crocodile's striking range is limited to a few yards, but it is only a matter of time before a leg or a muzzle comes within that distance. Suddenly one of the crocodiles explodes from the water, clamping its massive jaws on the first thing it contacts. Bluntly pointed teeth splinter through bone, locking fast on a startled wildebeest's leg. The impact throws the wildebeest onto its side in the shallows. If the crocodile had missed, it would have lunged again, but it would not have pursued its prey far onto the land—its domain is the water. The crocodile gives an enormous tug, dragging the wildebeest into deeper water. With hooves threshing and eyes popping, it lets out a terrified bellow. The next instant its head is jerked beneath the surface. Choking water sears down its throat and floods its lungs. The wildebeest struggles violently to get its head above the water but does not stand a chance against such brute strength.

When the struggling has ceased, the crocodile begins to dismember the carcass. It starts with one of the hind legs, delivering a series of ripping bites that sever tendons and smash through bones. Biting down onto the pulverized leg, the crocodile throws its body into a series of violent twists, rolling over and over in the water until the whole leg is torn free. With flicks of its head it throws the leg around in its mouth until it is properly positioned for swallowing, with the hoof pointing skyward. The jaws open and close, engaging the severed leg in the gullet, and the hoof disappears forever.

Often a second crocodile will assist by biting into the carcass, holding it steady while the other one rotates. Sometimes the two crocodiles will twist in opposite directions. Each individual eats what it bites off, without any aggression toward the other. Once the crocodile has gorged itself on wildebeest, it will go without food for several weeks. In contrast, a similar meal eaten by a lion would last only for a few days. Like other reptiles, crocodiles eat about two or three times their body mass in a year, compared with about twenty times for a lioness.

The powerful capturing and feeding activities of the crocodile last for several minutes and are probably fueled by anaerobic metabolism. Experiments have shown that crocodiles have a remarkable tolerance for the lactic acid that builds up in their blood, and this tolerance increases with body size. During these experiments, conducted by biologists in Australia, crocodiles of different body sizes were snared at the end of a length of rope and allowed to thresh around in the water until they were exhausted. The smallest individuals (under 1 kg, or 2 lb) were exhausted in about five minutes, whereas the largest ones (over 100 kg, or 220 lb) took over thirty minutes to become tired. Once exhausted, the crocodiles were incapable of any activity, for upward of several hours, while their bodies broke down the accumulated lactic acid. While the crocodiles were in this exhausted state, they were unresponsive—the experimenters could have tweaked their toothy snouts without fear of retribution. Contrast this slow recovery after

A Nile crocodile, with the severed leg of a wildebeest.

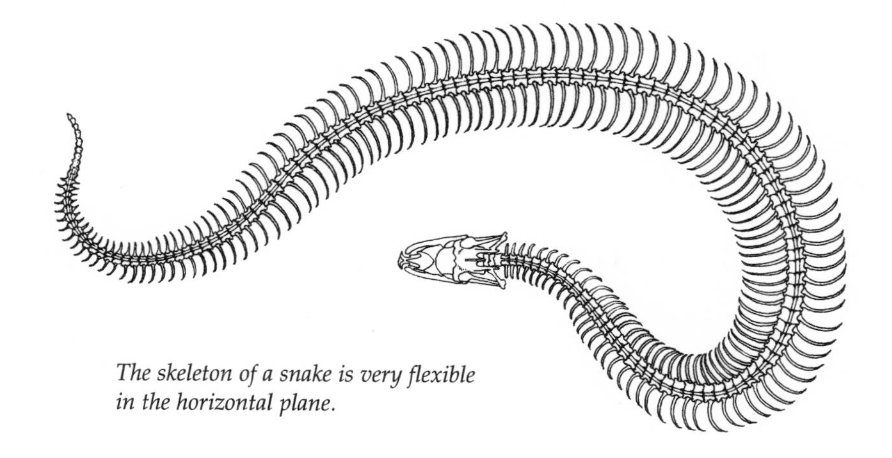

The skeleton of a snake is very flexible in the horizontal plane.

strenuous activity with the quick recovery of mammals. A cheetah's three-hundred-yard dash after a gazelle may leave it exhausted and panting, but it can still defend itself and is fully recovered within minutes.

Aside from its stealth, there is little sophistication in the hunting strategy of the crocodile; it is essentially a pair of powerful jaws mounted on the end of a rugged delivery system of immense strength. Snakes, in contrast, are delicately engineered killing machines with some of the most highly specialized weapons systems of all vertebrates. Like modern fighter aircraft, different types of snakes have different guidance and weapon systems, but the basic equipment is the same. The limbless body is elongate, and highly flexible in the horizontal plane. The skull is a lightly built structure of jointed struts that enable the two sides of the upper jaws to move independently. The two halves of the lower jaw are also jointed, and their articulation with the rest of the skull is very flexible, giving a gape of almost 180°. The needle-sharp teeth are long and slender and are curved backward toward the throat. The remarkable degree of flexibility in the skull, described as *cranial kinesis,* enables snakes to manipulate and swallow prey considerably larger than their own girth.

Many snakes are sit-and-wait predators, like crocodiles. They spend hours, even days, coiled up inconspicuously in a suitable location, waiting for an unsuspecting prey to come along. Some snakes actively search for prey, slithering

The skull of a snake is lightly built and remarkably flexible.

down burrows, climbing trees, and looking in other likely places. But they usually do not travel great distances in these forays, nor move at anything but a leisurely pace. There are some interesting exceptions, though. Cobras, for example, which become territorial during the breeding season, guard their domain against intrusion with considerable zeal. They raise their heads high off the ground, to peer over the top of the undergrowth, and human intruders have reportedly been chased over distances of many yards and bitten. [3]

Regardless of whether snakes go to the prey or the prey comes to them, their method of capture is the same. Other reptiles lunge at their prey with their entire body, but snakes, utilizing their elongate shape, strike out with the anterior region of their bodies. The purpose of the strike is to seize the prey, and in preparation, the neck is raised from the ground and thrown into a horizontal S-shaped bend. The strike is made by straightening out the curve, accelerating the head toward the target. Killing is achieved either by injecting venom into the prey through the fangs or by coiling the body around it, restricting respiratory movements and causing death by asphyxiation or cardiac arrest. Alternatively, killing is achieved simply by swallowing the prey after it has been struck. Snakes that kill by constriction tend to prey on birds and mammals rather than on other reptiles.

It has been estimated that over 100,000 people die from snakebites every year, [4] which explains why so much research has been conducted on snake venoms. Older textbooks usually classify venoms as *neurotoxic* or *hemolytic,* depending on their effect on the body; neurotoxic venoms interfere with nerve impulses, leading to paralysis, while hemolytic venoms break down blood and blood vessels, causing internal hemorrhaging. Although it is true that some snakebites do largely affect the nervous system, while others have greatest impact on the vascular system, this dichotomy is an oversimplification of a complex situation, because venoms frequently have both neurotoxic and hemolytic components. Venoms also contain a wide variety of other substances, including toxins that attack cardiac muscle, stopping the heart, and enzymes for breaking down proteins. The symptoms suffered by snakebite victims vary according to the species of snake, the amount of venom injected, and the site of the wound. There may be intense pain, local swelling, and necrosis (tissue death) at the site of the wound; lowered blood pressure, reduced heart rate, numbness, paralysis, breathing difficulties, internal bleeding, respiratory arrest, and cardiac failure. The prognosis for snakebite victims is not always grave. Few people die after being bitten by a rattlesnake, for example, the mortality rate being about 2 percent, but the bite of the black mamba (*Dendroaspis polylepis*) is fatal. Although snakebite victims often escape death, their injuries can be permanent and may result in the loss of limbs.

Snakebites are treated by giving injections of antibodies to specific venoms, referred to as *antivenin.* Antivenins are prepared by inoculating horses with small and increasing doses of snake venom, over a period of several weeks. During this time the horses build up resistance to the venom by producing specific antibodies to the venom in their blood. Several types of venom may be inoculated at the same time, to give a broad spectrum of antibodies for a number of snake species. Blood is then collected from the horses, and the serum containing the antibodies is extracted and purified. The venom used for the inoculations is obtained by "milking" live snakes. The snake is held with its open mouth pressed against the inside of a glass container, which causes the venom to be released from its fangs. Snake venoms are used not only for the production of antivenins but also for medical research. Among the many clinically interesting properties of venom is its ability to suppress tumors.

One unexpected use of snake venoms is in the detection of radioactive spills in the environment. A considerable part of the former Soviet Union is contaminated with radioactive isotopes, but information on the affected sites is not freely available to the inhabitants. These isotopes have made their way into living organism, including snakes, causing their venom to become radioactive. Snakes are collected for their venom throughout the territories of the old USSR and are sold throughout the world. This fact prompted two Russian scientists to publish an appeal for information in the international scientific journal *Nature.* They asked recipients of Russian snake venom to test it for radioactivity and to report back where the venom was collected, so they could inform the unwitting inhabitants of the contaminated areas.

A rattlesnake lies coiled beneath a fallen pine, as motionless as death. It has not moved for several hours. Unmoving eyes stare out on the world. Watching. Waiting. A young ground squirrel, inexperienced in the ways of life, approaches the pine and the promised harvest of seeds. The snake flicks out its forked tongue to sample the air and tastes the scent of rodent. The squirrel, its curiosity aroused by the sight of the snake, approaches within two yards and stops. The snake slowly slithers forward, sampling with its tongue as it closes the gap. The squirrel still does not move. The rattlesnake raises its head and makes a serpentine loop with its neck. Its movements are slow, cold, and calculating. Life hangs in the balance. The

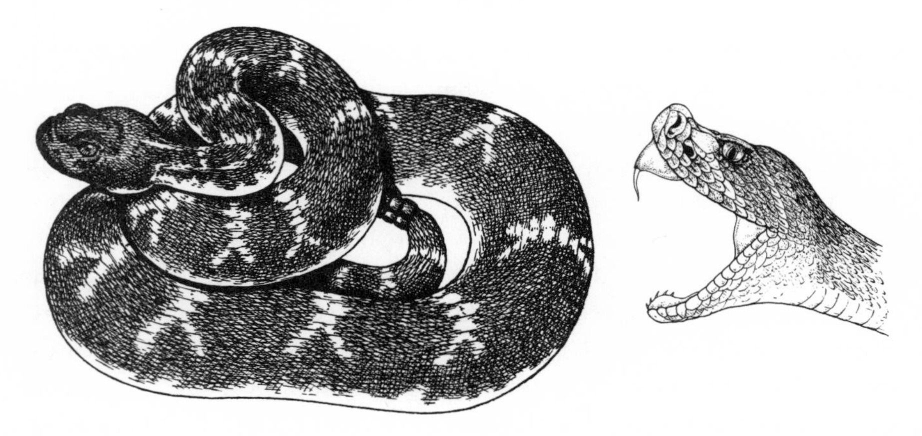

A rattlesnake prior to the strike. Note the S-shaped curve of the neck region. It is only after launching the strike that the mouth is opened and the fangs erected. The mouth shown here is not fully open.

next instant the squirrel is struck and immediately released. It all happens so fast that the action would have to be replayed in slow motion to appreciate the complexity of the sequence.

From the instant the snake launches the strike, it takes one-twentieth of a second for the head to accelerate toward the target. During that time the lower jaw gapes open, and the two sides of the upper jaw are depressed. The latter action causes a rotation in the bones attaching to the long pair of curved fangs that have been lying flat against the roof of the mouth. The fangs, which are hollow, now stand erect, in preparation for injecting the venom. The lower jaw makes contact with the squirrel's chest, and the upper jaw begins to close. During the next one-tenth of a second the snake arches its neck so that the downward thrust of the head follows a curved path similar to the curvature of the fangs. The fangs are driven deep into the squirrel's body, and the contracting jaw muscles press hard against the paired venom reservoirs, driving venom at high pressure through the fangs and into the wounds. The snake controls the amount of venom according to the size of its prey. The young ground squirrel receives a larger inoculation than would a small mouse. During the last twentieth of a second the mouth is flung open and the squirrel is released.

The sight of a snake killing a mammal, a young defenseless one at that, may not be a pleasant one, but we should not view the scene with sentimental eyes. Predators have to kill to eat, and do so without emotion.

Killing and being killed have nothing to do with assailant and victim, good and bad, only with survival.

The rattlesnake's venom may contain hemolytic agents, which begin breaking down the blood and blood vessels, causing internal bleeding. There are also neurotoxins, which will inhibit the squirrel's movements, eventually leading to paralysis. The venom also contains enzymes for breaking down proteins, and these enzymes begin digesting the squirrel's body from the inside. Its own heart assists the process by distributing the corrosive mixture to every part of its body. During these last agonizing minutes of the squirrel's life, it will scurry a few yards away from the snake. It is not in the snake's interest for the prey to get too far away; otherwise it might lose it. The snake could reduce this risk by injecting more venom, but that would leave less time for the digestive processes to work. The snake could simply hold the envenomated prey in its mouth until it died, and this sometimes happens. But rodents have sharp teeth, and rattlesnakes are therefore more likely to release them, to avoid the possibility of injury. The situation is entirely different when a bird or another reptile is struck. A bird would take flight on being released and would get so far away during its brief flight that the snake would never find it again. An envenomated reptile would also get far away, but for a different reason: venom acts far more slowly on ectotherms than on endotherms. Incidentally, venomous snakes appear to be resistant to injections of their own venom. They also have high tolerances to venoms of other species. This makes it difficult for snakes to kill other snakes, and species like the king snake (*Lampropeltis getulus*), which feed extensively on other snakes, therefore usually eat their prey alive. Seizing the prey as close to its head as possible, the king snake throws a few coils of its own body around its prey's body. Without releasing its bite, the king snake then works its mouth toward the prey's head. Rattlesnakes are their frequent prey, and king snakes appear to be quite immune to the bites they receive during this part of the procedure. Once the attacker's mouth reaches the rattlesnake's head, it engulfs it, and swallowing commences. By dragging the rattlesnake through the coils around its body at the same rate as it is eaten, the king snake can control its prey and keep its body taut. The sight of one snake devouring another is quite remarkable.

Most snakes have binocular vision, with at least 30–40° of overlap between the two eyes, and this gives them good depth perception, so vital in making a successful strike. But strikes sometimes go wrong. A snake may miss its prey altogether, in which event it may strike again. If a rattlesnake, or any other species with erectile fangs (the vipers), hits with only one of its fangs, it can often rectify the situation by swinging the other fang into position and driving it down into the prey while it is still in its mouth. In this way the prey receives the full dose of venom, so it will not go too far when it is released. The ability of vipers to move their long fangs independently makes them much more dangerous to handle than other snakes. My colleague at the museum, Bob Murphy, who works on snakes, was once bitten on the hand by a rattlesnake he was holding. The pain, which he likened to holding his hand on a hot stove, lasted for four days.

Snakes have a precise locating system for finding their envenomed prey—their forked tongue. The serpent's forked tongue, found in all snakes and many lizards, operates in conjunction with a pair of sensory organs located in the roof of the mouth. Each of these olfactory (smelling) organs, called *Jacobson's organs,* opens out into the mouth by a small hole. Chemicals in the environment are picked up on the tips of the tongue and transferred to the Jacobson's organs for analysis. For many years it was thought that the tongue tips delivered the samples directly to the organs, through the paired openings. However, more recent studies show that the tongue tips do not enter the organs. Instead, it seems that they deliver their chemicals to a pair of pads on the floor of the mouth, and these pads, in turn, deliver the chemicals to the sensory organs. The locating ability of the system is determined by how deeply the tongue is forked. Deeply forked tongues have long tines whose tips can be spread far apart during the sampling. This enables the detection of greater differences in chemical concentrations as the tongue samples the ground over which the dying prey traveled. In the same way, two radio receivers can triangulate on a transmitter, getting a more precise fix on its position if they are farther apart. By rapidly flicking the tongue as the snake follows the trail, it is able to determine whether the prey's scent is stronger on the left or on the right, thereby keeping on track.

When the snake locates the prey, it is carefully examined with the tongue. Once the snake is assured the prey is dead, it is grasped by the head and swallowed. Snakes have to swallow their food whole because they are unable to chew, their teeth being specialized for grasping. Since food cannot be mechanically broken up before being swallowed, the snake's digestive enzymes can only work on the outside of the prey's body. This explains the importance of the digestive enzymes injected into the prey with the venom. During swallowing, the snake's mouth works away as if it were made of rubber, and if the prey is very large, the process can last for over an hour. Such gastronomic marathons can be very tiring, and snakes often take rests during their meal.

Although rattlesnakes often sit and wait for a suitable meal to come their way, they will also actively seek out prey. The underground burrows of small mammals are a favorite hunting ground, the snake's sinuous body being ideally suited for slipping through narrow passages. Eyes are of little avail in those dark places, so the snake uses its tongue to locate prey. While the tongue is sufficiently sensitive for closing with the target, it lacks the necessary precision for launching the strike. For this purpose, the rattlesnake uses a separate system: a heat-seeking guidance system specifically designed for targeting warm-blooded prey in the dark. This system comprises a pair of pits, located between the nostril and the eye, which are sensitive to heat. Experiments with blindfolded snakes have shown that these thermal receptors are as accurate as the eyes at guiding the head toward the target. Snakes possessing this guidance system are referred to as pit vipers, to distinguish them from the vipers of Europe and Asia, which lack such a system. Some pythons independently evolved a similar system, which comprises a number of smaller pits. While snakes with heat sensors feed on endotherms, there are also lots of other snakes that lack the system but which nevertheless prey on birds and mammals.

Rattlesnakes often go hunting at night. Since they are ectothermic, their body heat is dependent on the sun, but their bodies cool down more slowly than they heat up. Snakes' ability to maximize solar warming and minimize cooling is achieved through several mechanisms. First, when they are sunning themselves, their hearts beat more rapidly, helping to distribute the warm blood from their skin to the rest of the body. They can also stretch out, exposing the maximum surface area to the sun. The reverse happens during the cooling phase: their heart rate is minimal, and they coil up to reduce their heat losses. Their narrow bodies also enable them to crawl into burrows and crevices, which also helps them conserve body heat. By these means, snakes are able to reduce their heat losses during the late afternoon, allowing them to become nocturnally active. Provided the night is not too cold, rattlesnakes are able to remain active, using their heat-seeking guidance system to prey upon warm-blooded animals.

Rattlesnakes are not as active as certain other species, like the racer (*Coluber constrictor*) and coachwhip (*Masticophis flagellum*). These fast, slenderly built snakes, which are native to North America, actively hunt for prey, relying on their speed to avoid becoming preyed upon themselves.

The coachwhip forages with its head held high off the ground, almost at right angles to its vertically oriented "neck." At the other end of the activity spectrum are heavily built and relatively slow moving snakes like the rosy boa (*Lichanura roseofusca*). This species, which is also native to North America, is a sit-and-wait predator. John Ruben, a comparative physiologist at Oregon State University, compared activity levels between the rosy boa, the western rattlesnake (*Crotalus viridis*), the racer, and the coachwhip. He did this by stimulating the snakes into vigorous exercise for a five-minute period. The rosy boas did not remain active for the entire five minutes, spending some of the time coiled up in a defensive ball. The rattlesnakes kept active but were completely exhausted at the end of the period, so much so that they would not even right themselves when placed on their backs. But the racers and coachwhips showed no signs of exhaustion, and were not completely exhausted even after another five minutes of exercise.

More than half of the energy used by the snakes during their forced activity was provided by anaerobic metabolism. But the racers and coachwhips had the advantage over the other species in generating much higher levels of aerobic energy—far higher than the other snakes could generate anaerobically. Part of the reason for the racers and coachwhips having so much stamina is that they have more complex lungs. These lungs are better supplied with blood vessels, allowing them to extract more oxygen from the air. Other snakes also appear to have lots of stamina, like the sidewinder rattlesnake (*Crotalus cerastes*). Sidewinders can travel well over 1/2 mile (1 km) during a single night's hunting. This is an impressive performance by reptilian standards but is insignificant compared with the 25-mile (40 km) hunting forays of African wild dogs.

During bursts of strenuous activity, a snake's total metabolic rate (aerobic plus anaerobic) can reach levels comparable to those of similarly sized mammals. However, its basal metabolic rate is only about one-tenth as high. A snake's food requirements are therefore modest, and it consumes only two or three times its body mass in a year. Its impact on its prey species is therefore minimal, in contrast to that of many mammalian predators.

Snakes cut down their energy costs during their long fasts by the remarkable strategy of reducing the size of certain of their organs, like the gut, lungs, heart, and kidneys. They can afford to do this because they make such small demands upon their bodies during the periods of inactivity. But the situation changes dramatically when it is time to feed again. The changes are most marked in snakes that have long fasting periods, and which take large meals. Such snakes, like pythons, may go for two or three months without feeding, sometimes even longer, but when they break their fast they have a large appetite. A meal weighing one-quarter of the snake's own body weight would be average, and one equaling its own weight would not be exceptional. To put this in human terms, a python's average repast is like our tucking into forty or fifty pounds of meat. The snake is understandably inactive after such a heavy meal, and it takes about fourteen days to digest it. During this time the organs that had become reduced in size increase, but not all at the same rate. Laboratory experiments on the Burmese python (*Python molurus*), for example, showed that the small intestine increased by nearly one-half of its fasting weight within six hours of feeding, while the lungs doubled in weight within fourteen days. These changes are accompanied by massive increases in metabolic rates. Within twenty-four hours of eating meals that were about one-quarter of its own body weight, the python's metabolic rate increased seventeenfold. But with meals weighing as much as the snake, the increase was by a staggering forty-five times. Such large increases in metabolic rates even surpass those of mammals when they are sprinting. What is more, mammals can only keep this up for short periods, whereas snakes can maintain their elevated metabolic rates for several days. There are three main reasons for the snake's remarkable metabolic performance during this time: it has to rebuild the organs that were reduced; it eats relatively larger meals than mammals; and its metabolic rate is very low to start with. There is a fourth possibility too: part of the increase in metabolic rate may be due to an increase in the snake's body temperature (a Q_{10} effect). [5] The energy required to fuel the increased metabolism comes from the snake's fat reserves. Snakes have large fat deposits in their bodies, hence the use of snake oil in herbal medicines.

Perhaps the most active of all reptilian predators is the Komodo dragon

The Komodo dragon.

(*Varanus komodoensis*). This large monitor lizard is endemic to Komodo and a few other islands in the Lesser Sunda group of Indonesia. While reaching lengths in excess of 10 feet (3 m), most individuals are much smaller, and the average size of the specimens that Walter Auffenberg measured during a yearlong field study was just under 6 feet (1.8 m). They are active scavengers, and in the absence of any other large scavengers on the islands, they have this food source more or less to themselves. They can detect the scent of rotting flesh from as far afield as a mile, using their forked tongue to sample the air. They roam over large distances in search of food, adult males traveling as far as 6 miles (10 km) in a day, though the average is closer to 1 mile. They usually move fairly slowly, about 3 mph (5 km/h), which is about half the speed of a foraging wild dog. But they can move much faster, and startled individuals have been clocked running at speeds of 9–11 mph (14–18.5 km/h). One large individual—6.5 feet (2 m)—was followed on a motorcycle at a speed of 9 mph (14 km/h), which it kept up for about two minutes. It is also reported that Komodo dragons can run at speeds approaching 20 mph (30 km/h) for distances of about 1/2 mile (1 km).[6]

Although they have a preference for carrion, they are also active hunters, adopting a sit-and-wait strategy like most other reptiles. Because they tend not to run very fast or very far, they have to get within about one yard of their prey to stand a good chance of capturing it. They hunt alone, preying on a wide range of animals, including cobras, vipers, rats, birds, monkeys, dogs, deer, and even water buffalo. Given that a water buffalo (*Bubalus bubalis*) can weigh over 1,000 pounds (over 450 kg), while the average weight of an adult Komodo dragon is about 100 pounds (45 kg), this is a remarkable feat. They are quite audacious and will occasionally enter villages to attack goats and cattle. There are also reports of their having killed and eaten villagers, though such incidents are probably rare.

While small individuals prey on rats and mice and other small animals, the large ones have a preference for wild deer and boar. A tactic they often use is to lie in ambush along game trails used by deer. If a deer approaches too close, it is seized, usually by a leg. The Komodo dragon has sharply serrated teeth, flattened from side to side like knives and specialized for cutting and slicing. Once the jaws have locked onto a limb, the dragon proceeds to saw and rip away by the powerful jerking movements of its head. Tendons and muscles are severed, causing lameness, and its actions are so vigorous that the prey is usually toppled to the ground. Choosing its moment, the voracious lizard exchanges its grip on the leg for the neck or abdomen. The neck is lacerated, rupturing major blood vessels and weakening the animal through blood loss. The abdomen is torn open, spilling out the intestines and other internal organs. Once the prey is dead, the predator begins feeding in earnest, usually starting with the viscera. The gut is vigorously shaken, to expel the partially digested food, before it is eaten. This delicacy is often followed by the diaphragm, then the lungs and the heart. The rest of the body is then cut up into manageable chunks by clamping the jaws tightly and rocking the head vigorously. Large quantities of flesh and offal are consumed in a short time, and a hungry adult can consume as much as 80 percent of its fasting body weight in a single meal. Like other lizards and snakes, Komodo dragons have very flexible skulls, which facilitate their feeding.

Other Komodo dragons are soon attracted to the site of the kill. A pack of hungry lizards makes short work of a carcass. Little is left to waste; they eat bones, hooves, even heads. But their digestive system is not as efficient as those of hyenas, and these hard parts pass through the gut undigested. As with other reptiles, their teeth are constantly being replaced, and are often swallowed along with the meal.

Komodo dragons are inactive at night and usually spend the hours of darkness resting in burrows. Because of their large body mass, they lose heat more slowly than smaller reptiles, and heat loss is further reduced by burrowing. As a consequence, their body temperature does not fall very far during the night, so they need to spend less time sunning themselves the following morning. Since the basal metabolic rate of reptiles increases with body temperature, the Komodo dragon's metabolic rate is higher at night than that of smaller reptiles. These large lizards also tend to forage far and wide during the daytime, but their food requirements are still modest compared with those of endotherms. It has been estimated that an adult consumes between three and four times its body mass per year, compared with about thirteen for a male lion and twenty for a lioness.

So far I have been concerned with reptiles as predators, but reptiles, and their cold-blooded allies the amphibians, possess a diverse array of defensive strategies. These range from the banal to the bizarre. Space limitations permit nothing more than a brief survey, starting with one of the most innocuous defenses—the tortoise's shell. Like a knight of old, the tortoise keeps its body encased in a protective armor, the efficacy of which has probably contributed to its long geologic history (tortoises date back to the

A horned lizard (also known as a horned toad).

Triassic period, some 230 million years ago) and to its longevity. When I was very young, my brother found a tortoise roaming around in a bombed-out house in wartime England. The tortoise appeared to have survived his human owners, the only damage suffered being a burn on the top of his shell. He lived until just a few years ago, surviving many more scrapes along the way. At one time we had a German shepherd dog who used to amuse himself by attacking the tortoise. The tortoise responded by withdrawing into his shell. Although I often had to go and find the tortoise after the dog had buried him, he never suffered any injuries. Many other reptiles, like horned lizards (*Phrynosoma*), are armored with long spines which serve a similar role.

Equally innocuous as the tortoise's shell are the various artifices that some reptiles use to make themselves look more dangerous than they really are. The frillneck lizard of Australia (*Chlamydosaurus kingii*), for example, has an enormous cloak of skin around its head and shoulders that it erects to make itself look large and terrifying. The cobra's hooded display is also defensive; so too is the way that many snakes, like the boomslang (*Dispholidus typus*), puff themselves up to intimidate predators. The hissing of reptiles, by no means confined to snakes, is also used to frighten off potential attackers. The effect can be quite startling, as I discovered when I was hissed at by a giant tortoise in Galápagos. The North American bull snake (*Pituophis catenifer*) makes a loud blowing noise, which has been likened to the bellowing of a bull, and the rattlesnake rattles its tail as a defensive mechanism too. The fact that such dangerous animals as rattlesnakes and cobras go to some lengths to frighten off would-be attackers underscores the vulnerability of their frail bodies. Coyotes, for example, regularly kill rattlesnakes simply by grabbing them and shaking them vigorously, which breaks their backs.

Snakes, as we have seen, swallow their prey headfirst. A defensive strategy adopted by some lizards is to hold their tails in their mouths, mak-

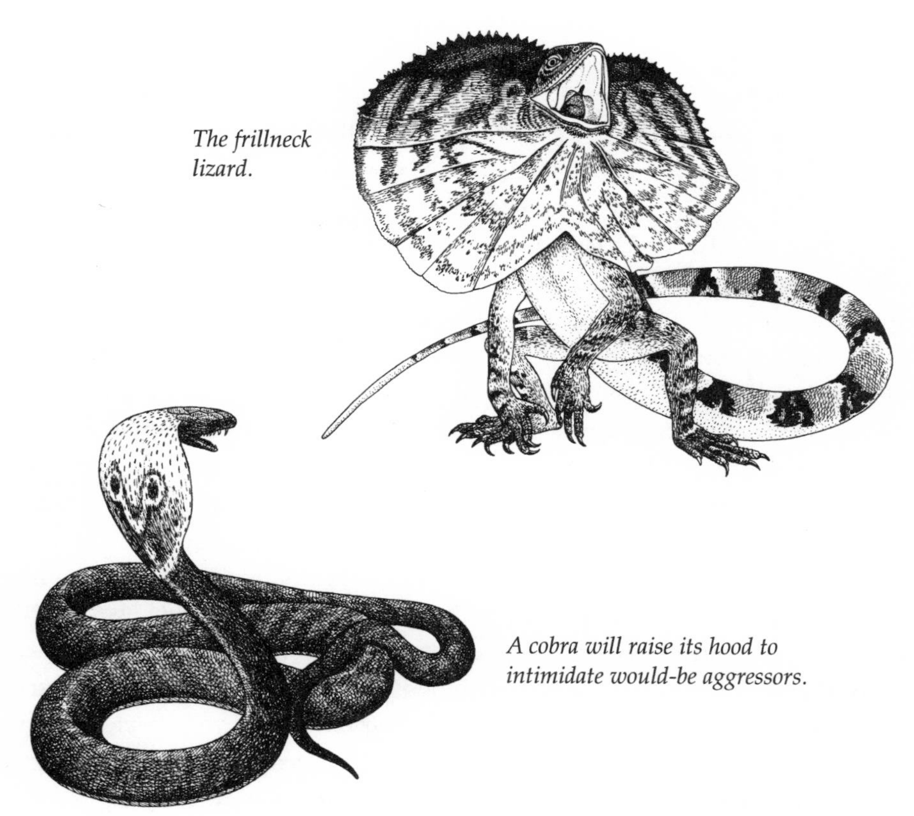

The frillneck lizard.

A cobra will raise its hood to intimidate would-be aggressors.

A rattlesnake in a defensive posture, rattle shaking.

Some lizards seize their own tails to prevent themselves from being swallowed.

ing it impossible for them to be swallowed. Many salamanders exude a sticky mucus from their skin, and this can actually glue them to the body of an attacking snake, completely thwarting the snake's efforts to swallow them. Many lizards, and some snakes, are able to shed their tails when threatened or attacked. This mechanism, called *autotomy,* is an active process involving planes of weakness in the tail vertebrae that are split apart by muscle contractions. If the lizard's tail is being held at the time it breaks off, the owner can flee, leaving its attacker with a partial meal to distract its attention. If the predator has not yet seized the lizard, the sudden appearance of a squirming tail is often a sufficient distraction to allow the lizard to escape. A new tail is soon regenerated, and it is not uncommon to find a lizard in the wild with a scar across its tail where the previous one was shed.

The horned lizard *Phrynosoma cornutum,* which is well endowed with defensive spines, has the peculiar ability to squirt blood from its eyes. The reaction is primarily elicited by canids, and by adjusting the position of its head, the lizard can aim the paired blood jets at the attacker's eyes and face. The blood has no harmful effect on the predator's eyes, its effectiveness probably being in its shock value. Several species of snakes, including cobras, spit venom from their fangs. If this reaches the eyes, it can cause pain, swelling, and temporary or even permanent blindness. Some snakes, including the rosy boa, exude an evil-smelling fluid from their cloacae when threatened. (The cloaca is a common opening through which the gut and urinogenital system discharge.) Others, like the harmless English grass snake (*Natrix natrix*), are able to secrete an irritating fluid from their skin. But when it comes to noxious skin secretions, the amphibians are without equal.

The moisture in the skin of amphibians comes from watery secretions produced by numerous small glands in the skin. Moist surfaces are natural harboring grounds for bacteria, and some of the chemicals secreted by the glands may serve to kill bacteria, thus preventing infection. Some of these chemicals are distasteful, discouraging predators from touching them. Others range from being mildly toxic to highly poisonous. The extreme example of the latter are the highly poisonous tree frogs of the South American rain forests, which native hunters use to

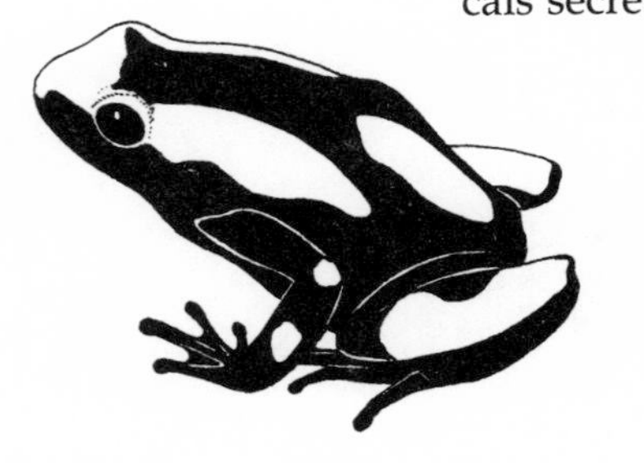

A poisonous South American tree frog.

poison their arrows. These little frogs are brightly colored, which serves as a warning to potential predators to stay away—like the black and yellow stripes of the wasp.

An extreme variant of poison skin glands is provided by a salamander from Japan (*Echinotriton andersoni*). This brown-and-orange salamander has a series of nine to twelve large warts along either side of its body. The skin covering these warts has glands that produce a toxin, and there is a rib in the center of each wart. What is more, these ribs are sharply pointed and stop just short of the skin. When threatened by a predator, the salamander begins posturing, raising its tail and straightening its back legs to reveal the bright orange underside of its tail. The ribs are then raised, puncturing the warts, the objective being to inoculate the predator with the toxin.

In spite of their toxicity, some amphibians are still eaten by certain predators. This raises the question of whether these predators have evolved some degree of immunity to the toxins. To answer this question, a study was made of some toxic newts and the garter snakes that fed on them. It was found that the snakes that fed on the newts had higher tolerances to their toxins than garter snakes living in areas where the newts did not occur. Sometimes snakes that lack immunity swallow toxic newts. It would appear from this that the newt's toxicity is of no selective advantage to it, because once the newt has been swallowed it cannot pass its favorable defensive genes on to the next generation. This makes it difficult to see how toxicity could have evolved in the first place. However, being swallowed by a snake does not necessarily kill a newt. Toxic newts have been seen crawling out of a dead snake's mouth after their poisons have done their work—a cold-blooded escape from a cold-blooded death.

3

Death at Sea

Sable Island is a low sandbar 100 miles (160 km) off Canada's rugged east coast. During the last ice age it was part of the land, but it is now the highest point of a submerged bank—part of the continental shelf that skirts North America. The twenty-four-mile stretch of sand dunes and beach grass is an incongruous sight so far out into the Atlantic, where ocean swells can form hills almost as substantive as the land itself. The shelving seabed, acting in concert with heavy winds, can raise some treacherous seas. And the fact that Sable Island lies so close to the transatlantic shipping lanes made it one of the most dangerous threats to navigation since the early days of sail. Shipwrecks once littered the area, earning the island the epithet Graveyard of the Atlantic. The loss of life and property became so serious that a permanent settlement was set up in 1801—the Sable Island Humane Establishment—charged with saving lives and salvaging wrecks. The work of the mission continued well into the present century, but with the advent of radar, shipping losses declined, the last wreck being recorded in 1947.

This narrow strip of land, barely 1 mile (1.6 km) wide, is home to seabirds and seals, and to a population of wild horses—descended from those introduced by earlier settlers. It is also the home of an unresolved series of killings.

Zoe Lucas, a wildlife biologist, has been studying the biology and ecology of Sable Island since 1974. But during the last few years she began

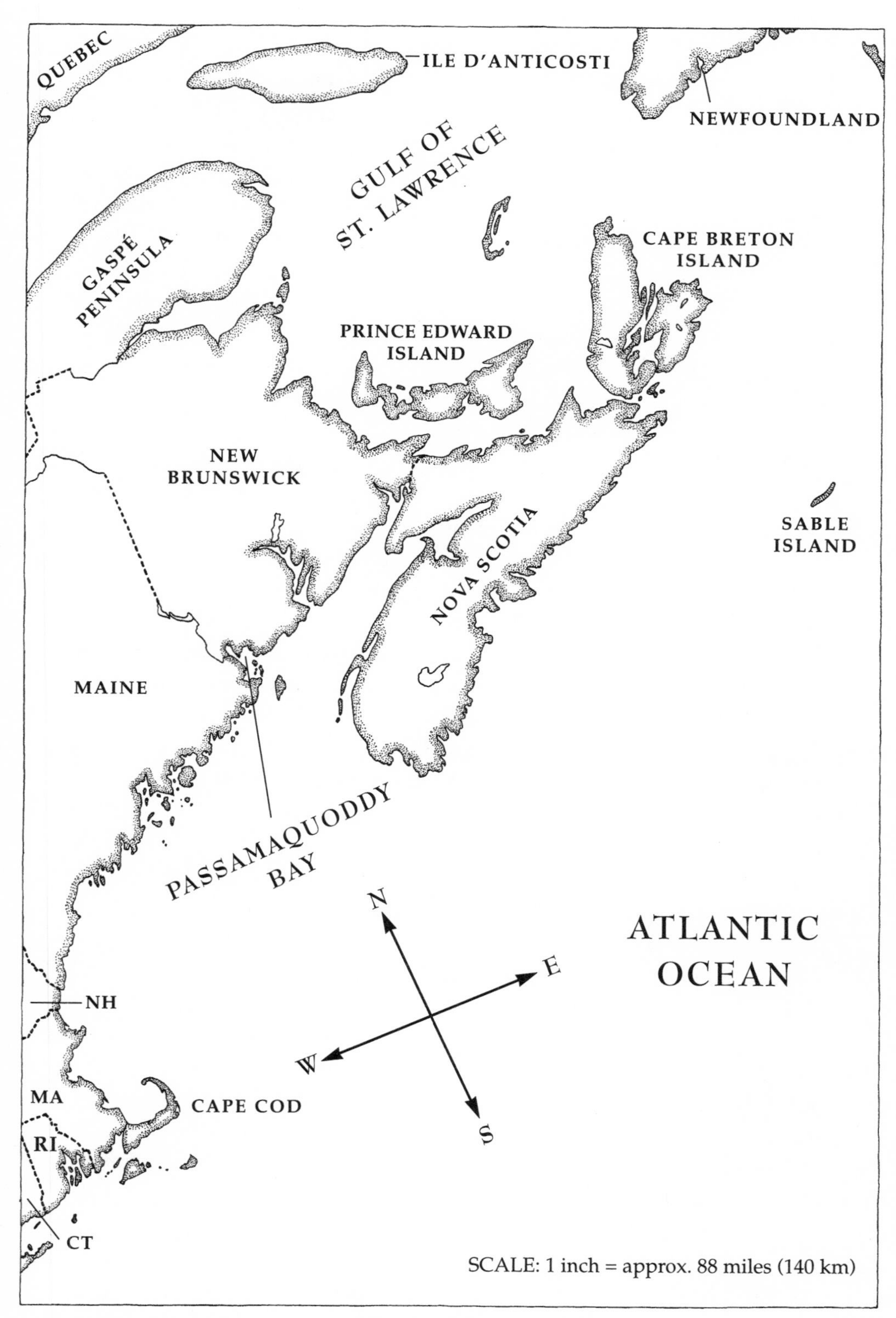

Sable Island lies about one hundred miles off the coast of Nova Scotia.

noticing something very unusual. Seals were dying, by the hundreds, under most unusual circumstances, and their strangely mutilated bodies were washing up on shore all around the island. Four species of seals were involved, but most of their injuries were the same: Their bodies were lacerated by a deep spiral cut, usually confined to the head and trunk, that encircled the body. The cut was clean and deep, right through the skin and blubber, which usually separated from the underlying flesh like the skin from an orange. What could have inflicted such mechanically precise wounds? A ship's propeller? The impeller blades in a water inlet of some power plant? Both of these mechanical devices could be ruled out; there were too few boats and no power plants within two hundred miles. Could another animal have inflicted such wounds? Sometimes the hind portion of a seal washes ashore, as if it had been bitten in half by a large shark, like a great white (*Carcharodon carcharias*). However, Lucas believes that the hind portions represent the remnant of a spirally injured carcass, rather than severed ones. Once a body has been lacerated, there is nothing to hold it together, and the front portion therefore breaks up in the surf and decomposes, leaving the intact tail section.

If these unusual deaths had occurred in the terrestrial environment, the mystery would have been resolved in no time. This is simply because most of what happens in the sea is unseen by human eyes. We catch only the occasional glimpse of life in the ocean, as from the deck of a ship or some vantage point on the shore. And even when we put on scuba gear and dive beneath the surface, we see only a fraction of the upper layers of the sea. It is therefore not surprising that so much of what happens in the ocean is a complete mystery to us. All we can do is catch a glimpse here, and scrap of evidence there, and try to put the pieces of the puzzle together.

Our next stop is some 350 miles (550 km) southwest of Sable Island, at a point just off Passamaquoddy Bay, still on Canada's eastern seaboard.

The sea is gray. The sky is gray. The heavy mist will clear by midmorning to reveal a fleet of rocky islands moored beneath a summer sky. But for the time being the world is a cold and damp place that has lost the sun. There is no wind to ruffle the sea, no noise of surf against the nearby shore. The waiting silence is broken only by the periodic wailing of a distant foghorn. The sea's movements are sluggish, like heavy oil. A large frond of bladder wrack, plucked from the rocks during a forgotten storm, rises and falls with the sullen swell. And beneath the surface, a lighter shade of gray can just be discerned. To the discriminating eye of a seabird, the in-

The harbor porpoise is one of the smallest cetaceans.

The blue whale is the largest cetacean and the largest animal to live on Earth.

*distinct shape might be recognized as a fish, but the birds are not fishing this morning. They must wait until the mist clears. Yet the poor visibility is no impediment to the dark shape looming out of the depths. Guided by a series of sonar clicks audible to the human ear, the 5-foot-long (1.5 m) torpedo homes in on its target. Seconds later it seizes the slippery fish in its sharply pointed teeth and swallows it whole. The dark shape, moving so smoothly through the water, can now be identified as a harbor porpoise (*Phocoena phocoena*).*

Rarely exceeding 6 feet (1.8 m) in length and 200 pounds (90 kg) in weight, the harbor porpoise is one of the smallest cetaceans. The cetaceans—porpoises, dolphins, and whales—are the most highly modified of all the marine mammals. Their streamlined bodies are covered in smooth skin, like a neoprene diving suit, and insulated with blubber. They have a single dorsal fin, paired forefins but no hindfins, and a horizontal tail with paired flukes that generates the propulsive thrust of swimming. They are divided into two main groups: the odontocetes, or toothed whales; and the mysticetes, or whalebone whales. The toothed whales all have teeth and feed on large food items like fishes, squid, and other marine mammals. They include porpoises, dolphins, killer whales, and sperm whales. The mysticetes are all without teeth, having instead horny plates, called *baleen*, that hang on either side of the mouth, like curtains, to sieve small food items, mainly plankton, from the water. They are all large. The smallest mysticete, the pygmy right whale (*Caperea marginata*), is about 16 feet (5 m) long. The biggest, the blue whale (*Balaenoptera musculus*), reaches lengths of up to 100 feet (31 m) and is the largest animal ever to have lived on Earth. *Sonar echolocation,* used both for navigation and for locating prey, is a feature of odontocetes. Pulses of sounds—some as low as 1–5 kilohertz (kHz), or 1,000–5,000 cycles per second—are sent out through the water, and the rebounding echo is picked up by the ears and analyzed by the brain. The brain computes the bearing of the target, its distance downrange, and possibly also provides some information on the nature of the target. This complex mechanism is far from well understood. The sound is generated in a system of ducts that run through a spherical structure, aptly named the *melon,* that lies on top of the skull, between the tip of the snout and the eyes. The melon is filled with fatty material, which may act as a lens to focus the sound into a narrow directional beam.

Being warm-blooded, like other mammals, the harbor porpoise has much higher food requirements than the cold-blooded fishes upon which it preys. The sea around Canada's eastern coast is always cool, surface temperatures seldom exceeding about 16°C (60°F) in the summer. But the porpoise's blubber insulates it against the cold. The blubber is also important as a food storage depot. This is especially important in the mysticete whales, which fast for many weeks during their long migrations, and also during times of low plankton production. Their blubber is correspondingly thicker and is much sought after by certain predators as a high-energy food source.

The foghorn's mournful cry tolls out through the gloom—then all is cold gray silence again. The sea looks dark and uninviting. The porpoise breaks surface with a rasp of exhaled air, snatching a fresh breath just before its blowhole disappears beneath the surface again. Arched back follows bulbous head in the same fluid motion, slipping beneath the surface as smoothly as silk. The porpoise, swimming at about 3 mph (5 km/h), has already located another target, just a few feet below the surface. But before it has grasped the fish, its own body is seized from below and bitten in half. The back end of the porpoise is immediately swallowed, and the momentum imparted by the unseen predator propels the head and torso to the surface in a dark cloud of blood. The severed carcass bobs in an intestinal flotsam for several moments before slipping beneath the surface for the last time. A gull materializes from the fog and begins feasting on the unexpected bounty.

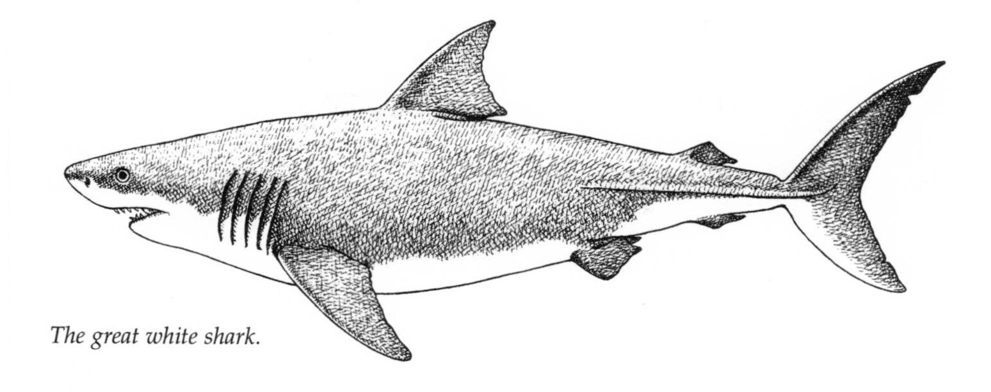

The great white shark.

The unseen predator, a 17-foot-long (5.2 m) white shark, is not normally associated with cold Canadian waters, but it may be an important predator on seals and porpoises in this region. The shark has two other porpoises in its belly, both with their tails lopped off, which shows that they were attacked from behind. Chopping off the tail is a convenient way of incapacitating a strong swimmer like a porpoise, enabling the shark to feed at leisure.

The white shark, of *Jaws* fame, enjoys notoriety as a man-eater, especially in warmer parts of the world like Australia. While its reputation is well deserved, the number of human fatalities is small, and most attacks probably occur because people are mistaken for the shark's usual prey of seals and sea lions. The attacks are usually made by large individuals—animals over 6 feet (1.8 m) in length—because smaller ones feed mostly on fishes. Not all white-shark attacks on humans are fatal. Sometimes a leg or other part of the body is seized and subsequently released. This seems to happen more frequently if the victim strikes out at the shark's head or eyes, both of which are sensitive to touch. But merely being grasped and released again can still inflict a serious wound, with extensive blood loss. This is because the triangular teeth, which are characteristic of the species, have sharply serrated edges, like a steak knife. The teeth are numerous and are constantly being replaced. The new ones form on the inside of the mouth and jaw, moving slowly forward as they grow, as if they were on a conveyor belt. The teeth are specialized for slicing through flesh, leaving a large circular bite mark. But they can also cut through tendon and bone by means of a sawing action, which sharks achieve by shaking their victim in their jaws. The 13-foot (4 m) white shark that attacked José Miranda off the Chilean coast in 1980 had no difficulty removing his left arm and shoulder, nor in decapitating him.

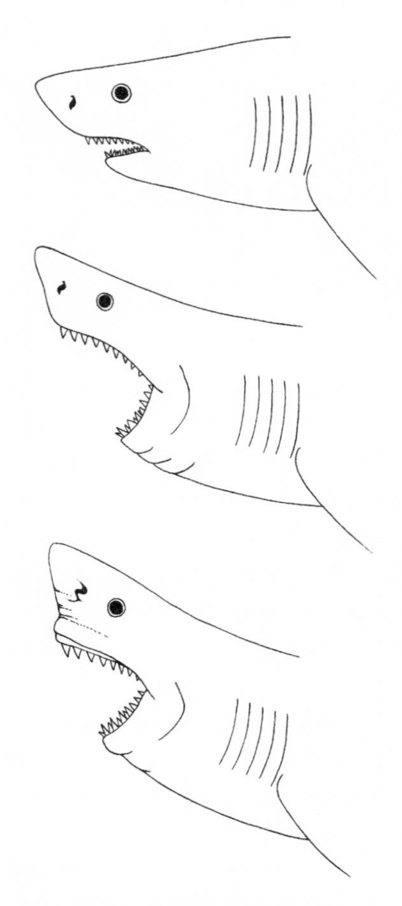

Prior to biting, the white shark opens its mouth and lowers its upper jaw, thereby elevating its upper teeth.

Although there are numerous accounts of the exploits of the white shark, information on its natural history has been less forthcoming. The biting mechanism of the jaws has been investigated by suspending pieces of meat in the sea from boats and filming the sharks that come to feed. One of the most extensive investigations of this kind was made off Dangerous Reef, in South Australia, by American zoologists Timothy Tricas and John McCosker. Analysis of the film showed that the white shark lifts its snout and head prior to biting, and drops its lower jaw. The upper jaw is then displaced downward and forward, elevating and exposing the upper teeth in a malevolent grimace. These preparations for biting increase the bite force and effectiveness of the teeth. One of the best places for seeing white sharks in action, attacking their natural mammalian prey, is off the California coast.

Since the late 1980s Peter Klimley, a marine behaviorist at the University of California, has been studying the behavior of white sharks around the South Farallon Islands, 30 miles (50 km) west of San Francisco. This is a prime location because the rocky islands are breeding grounds for sea lions and seals—collectively termed pinnipeds (meaning "winged feet" in reference to their flippers). There is also a good vantage point on a prominent hill for observations, and a videotape record has been made of daylight attacks for the period 1987–92. During that time 131 attacks were recorded and analyzed. The sharks frequently attacked the pinnipeds near to shore, often close to where they hauled out of the water or slipped into the sea. Seals, including the massive northern elephant seal (*Mirounga angustirostris*), were attacked more often than sea lions (California sea lion, *Zalophus californianus*, and Steller sea lion, *Eumetopias jubatus*). The sharks ranged in length up to about 16 feet (5 m) and weighed up to about 1 ton. Elephant seals reach lengths of 18 feet (5.5 m) and weigh up to about 3 tons. The sharks used different tactics on the seals than they did on the sea lions, and this was thought to relate to differences in their anatomy. Seals have strong hind-flippers that are directed backward and used solely for swimming, with short fore-flippers. Most species do not use their flippers on land and move by a laborious undulation of the body, like a caterpillar. Sea lions, in contrast, have long fore-flippers, used for swimming and walking, and hind-flippers that can be turned forward beneath the body, to support them when moving on land.

The first indication of a shark attack on a seal was usually the appearance of a patch of blood on the surface, with no sign of either the attacker or its prey. Then the seal would surface, followed by the shark, usually close beside it. Alternatively, the appearance of a bloodstain on the water would be followed by the sight of a shark swimming along the surface, using exaggerated beats of its tail. The modified swimming pattern was to compensate for the extra drag and extra weight of the seal that the shark was carrying. The seal might not be visible, though, because a shark's jaws are underwater when it is swimming on the surface.

The shark usually launches its attack from below, and it seems likely that it cruises along, scanning the sea above for a suitable target, silhouetted against the light. The white shark has good eyesight, the retina of the eye having a high density of cone cells, which are associated with daytime vision. Klimley reconstructed the most likely sequence of events in a white-shark attack on a seal. Once a target is spotted, the shark ascends, seizes the prey in its jaws, and carries it beneath the surface. The initial bite causes profuse bleeding, resulting in the large bloodstain on the surface. Sometimes the seal floats to the surface shortly after the attack, or there may be a long delay before it reappears—presumably because the shark still has the seal in its jaws. In any event, by the time the seal surfaces it is already dead, from the massive loss of blood. The shark usually follows its prey to the surface and may continue feeding. Sometimes the shark postpones feeding for as long as two hours.

Sea lions are also attacked from below, but the shark usually charges them with so much force that they are knocked out of the water. The first sign of an attack is therefore when a sea lion and shark burst from the water in an explosion of spray. The shark then dives, carrying the sea lion with it, but it might bite down much sooner than with a seal, releasing the sea lion while it is still alive. This earlier release may be in response to being struck by the sea lion's powerful forefins. The badly wounded sea lion writhes slowly along the surface, in a desperate attempt to escape. But it does not get far before the shark reappears, seizes it, and carries it below the surface for the second time. As in the case of the seal, the sea lion's body eventually floats to the surface. The shark surfaces once more and continues feeding, either right away or some time later. The common element in these attacks on pinnipeds is that the prey is seized, released, then seized again, death occurring through exsanguination (blood loss).

Humans are such cursory visitors to the sea that they are not included as part of a shark's normal diet. But a human who happens to be in the wrong place at the wrong time can easily be targeted as a food item. The fatal attack on the Chilean diver, for example, occurred close to where sea lions hauled out of the sea. Several days before the attack, local divers had noticed sea lion parts on the sea floor, but they did not heed the warning that this was a feeding ground for sharks. The shark's initial attack on the diver was fatal, but this is not always the case. If a victim can be hauled from the water before the shark makes its follow-up attack, that person can be saved. Scuba diving in the company of others obviously increases the chances of removing a victim from the water, in the event of a shark attack.

It has been suggested that a scuba diver's black neoprene wet suit, and fins, by enhancing his resemblance to a marine mammal, may be a contributing factor to shark attacks. Another way in which humans might be mistaken for pinnipeds is when they are floating at the surface on a surfboard. Seen from below, the silhouette of a surfboard, with arms and legs trailing in the water, looks very much like a pinniped. Sharks certainly do attack surfboarders, and it has been proposed that they have a search image which a surfer's silhouette against the sky resembles. The idea is certainly plausible, but it may not be true. The point is that sharks attack a wide variety of floating objects, from white buoys and bright green kayaks, to black Zodiacs and yellow surfboards. Since any object viewed from below against a bright sky appears in silhouette, color is probably immaterial. The wide variety of objects that evoke attacks suggests the same is true for shape. Therefore, any suitably sized floating object is liable to be attacked by a shark, regardless of its color or shape. This is especially so if the object is moving, and if it is in the vicinity of the shark's natural prey. So a human, in the wrong place at the wrong time, is probably equally vulnerable to attack, whether swimming, lying on a surfboard, or scuba diving.

Like any other fishing story, there have been many exaggerated claims of the sizes attained by the white shark. Based on measurements of some jaws in London's Natural History Museum, a Victorian author estimated a body length of 36.5 feet (11.1 m). This figure found its way into the textbooks, but when the specimen was reexamined in the early 1970s, it was realized that a mistake had been made and that the estimated length was closer to 16.5 feet (5 m). The largest white shark for which there appears to be reliable measurements was a 21-footer (6.4 m), caught off the Cuban coast, but this record has been questioned too. A shark of this length would weigh about 3.5 tons (3,500 kg) and have 2-inch-long teeth (5 cm). Large sharks like this would be capable of attacking most other animals in the sea. White sharks have been seen feeding on dead whales, but it is unlikely that even the largest individual would be able to kill a large whale on its own, and there is no evidence that they ever hunt cooperatively.

The occurrence of white sharks feeding on floating whales provides a unique opportunity to study them under natural conditions. In the summer of 1979 a 50-foot-long (15.3 m) fin whale carcass was seen floating off Long Island, New York, attended by several hungry white sharks. The carcass, which was in a state of advanced decomposition, was trailing an oil slick several kilometers long, attracting sharks from far and wide. Although other shark species are abundant in these waters during the summer, they kept well clear of the whale carcass and the white sharks, maintaining an exclusion zone over 3 miles (5 km) in diameter. The carcass was kept under observation for just over a week, during which time it was visited by up to nine white sharks. At least one of their number followed the drifting carcass, feeding from time to time.

The whale carcass was floating low in the water, and most of the sharks fed on the submerged parts. They were primarily interested in the blubber, and typically fed by rolling upside down and biting into the carcass. Then, with much thrashing of the tail, they rolled upright, ripping out a mouthful of blubber. There were never more than two sharks around the carcass at any one time, because of the aggressive rivalry among them. On one occasion a 10- to 13-foot (3–4 m) shark approached the whale, then quickly changed direction, a few seconds before a 16- to 20-foot (5–6 m) male appeared. The smaller shark already bore tooth marks on its skin, probably the result of a previous encounter with a larger shark. Whale carcasses, both floating and submerged, may provide an important food source for white sharks, especially outside of the pinniped breeding season when this resource is unavailable to them.

One of the biologists took the opportunity of attaching a tracking device to a 15-foot (4.6 m) shark, which was tracked by radio for three and a half days. During this time the shark covered a distance of 120 miles (190 km), at an average speed of 2 mph (3.2 km/h). This may seem slow for such a speedy-looking fish but is similar to results obtained for other large and seemingly fast fishes like marlins, tunas, swordfish, and mako sharks. Although these pelagic (of the open ocean) fishes may be capable of short bursts of speed, they spend most of their time cruising fairly slowly, a strategy for reducing energy costs. The tracking device recorded both the water temperature and the shark's body temperature, together with the water pressure, from which water depth could be deduced. Although the shark occasionally dived into colder waters over 150 feet (45 m) deep, it spent most of its time cruising in the upper 50 feet (15 m), where water temperatures were much higher.

The body temperatures of most fishes are the same as the surrounding water, because the heat generated by their muscles is so rapidly lost through the skin and especially through the gills. However, some fishes, like tunas and certain sharks, including the white shark, retain some of this heat by a heat-exchange mechanism. Warm, oxygen-depleted blood on its way to the gills passes through a special network of fine blood vessels lying next to a similar network carrying blood that has already circulated through the gills. The warm oxygen-depleted blood en route to the gills therefore gives up its heat to the cold oxygenated blood returning from the gills, so that the heat is not lost to the water. One of the advantages of elevating the body temperature is that more power can be generated by the muscles, a fact that is well known to athletes. Sharks have not gone very far in this direction; their body temperatures fluctuate with sea temperatures and are only a few degrees higher. The white shark's body temperature was only about 5 C° (9 F°) higher than the surrounding water.[1] However, large tunas, which reach lengths of up to about 10 feet (3 m) and weights of 1,400 pounds (650 kg), can maintain constant body temperatures of about 35°C (95°F).

The swordfish, a close relative of the tuna, does not keep its body temperature elevated but has a brain heater to keep its brain warm. This reddish brown structure lies between the eyes and is actually a pair of modified eye muscles. The ability to contract has been lost, but the muscle tissue has such a high metabolic rate that it generates sufficient heat to keep the eyes and brain at about 10–14 C° (18–25 F°) above ambient temperatures. The swordfish is therefore able to maintain optimal brain function when it dives from the warm surface layers to the cold ocean depths. This large fish, which reaches lengths in excess of 12 feet (3.7 m), is an active predator. It uses its sharp sword to slash at and incapacitate its prey. Although we know little about them, we do know that swordfishes appear to have an aggressive streak; there are numerous accounts of their ramming ships, and whales, with their long swords. The swords frequently become permanently embedded in the objects they strike, often snapping off, and it is not rare for a fisherman to land a swordfish that has a broken sword. Some of the rammings may be accidental, the outcome of a sharp object, traveling with considerable momentum, striking a large obstacle that happens to be in its path. Others are unquestionably deliberate and might have to do with territorial behavior.

From the body temperature data obtained from the white shark, the researchers estimated that it had a low metabolic rate, and that a single meal would last an individual for more than a month. This certainly makes good sense, and we should dismiss the notion that sharks are forever on the prowl for something to eat. Having a low metabolic rate, though, does not restrict sharks to the same stamina straitjacket that typifies reptiles. One of the reasons for this is that the energetic costs for aquatic vertebrates are much lower than for terrestrial ones—only about one-tenth as high. Another reason is that as fishes swim faster, they can increase the flow of water over their gills by opening their mouths, using a ram-jet mechanism, thereby raising the rate of oxygen uptake. Sharks can therefore pursue mammals with the stamina of a lion chasing a zebra. But they do not live on an exclusive diet of pinnipeds and dead whales. Like most other predators, they are opportunistic, eating anything that happens their way. They probably include a good deal of fish in their diet, and accounts of the remarkable things that have been found in their stomachs are legion and often greatly exaggerated. One of these questionable accounts reports finding an old automobile license plate, an unopened can of paint, a roll of tar paper, and a woman's head, all inside one shark!

Although the white shark record of 21 feet (6.4 m) has been questioned, it seems likely that such large individuals do occur. Measurements of the jaw marks inflicted on a whale carcass found drifting off the coast of western Australia, for example, gave an estimated length of about 26 feet (8 m). Though we should not put too much reliance on estimates like these, they do indicate that large sharks, probably in excess of 20 feet (6 m), probably do exist. Sharks much larger than this existed in the distant past, and estimates based on the size of fossil teeth indicate lengths of about 43 feet (13 m). A shark of this size might have been capable of attacking a large whale on its own, so perhaps sharks' modern habit of feeding on whale carcasses represents a remnant behavior from the past.

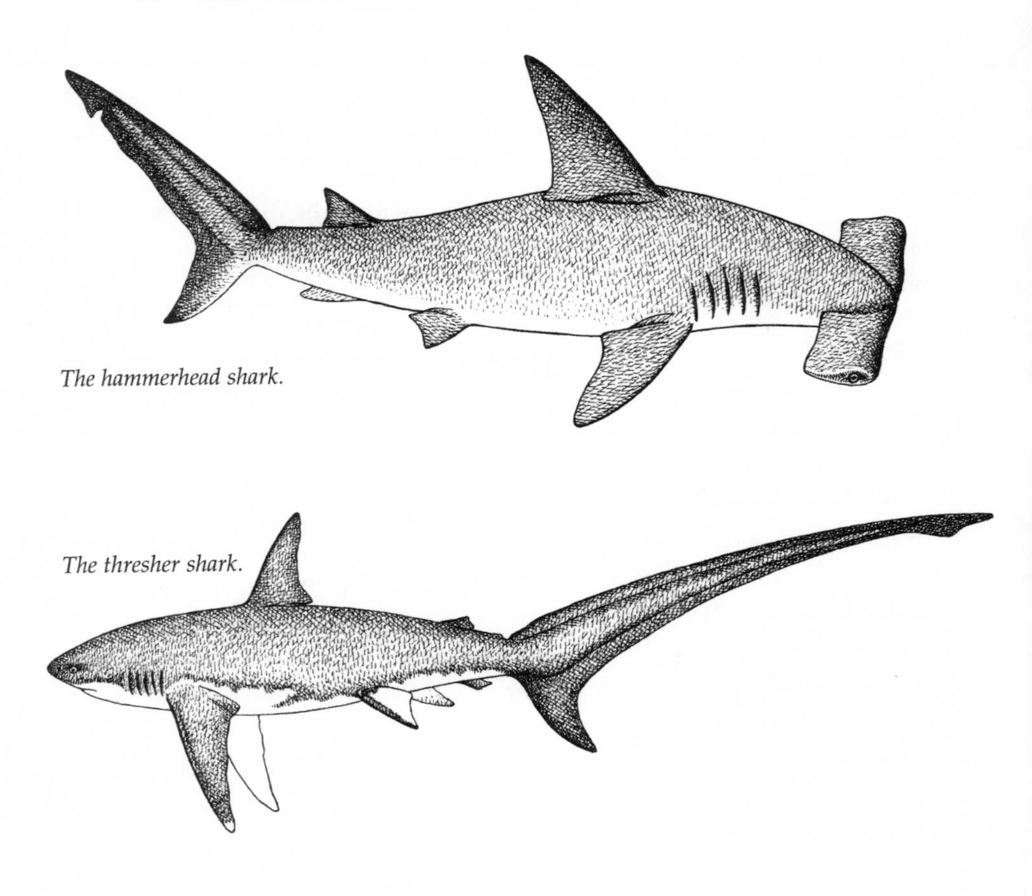

The hammerhead shark.

The thresher shark.

Before leaving the subject of sharks and moving on to killer whales, I would like to mention the hammerhead sharks (*Sphyrna* and *Eusphyra*) and the thresher shark (*Alopias*).

The hammerhead shark, of which there are several species, is unusual for the lateral expansion of its head into a pair of foils, like stubby wings, with the eyes placed at the ends. One possible function is that they act like hydrofoils to produce lift, enhancing maneuverability in the vertical plane. The expansion of the head also increases the distance between the olfactory organs, which may improve the shark's ability to follow scent trails in the water. While these possible functions are conjectural, some investigators working in the Bahamas had the rare opportunity of seeing hammerhead sharks using their heads to attack their prey, in this instance stingrays. Stingrays swim through the water by flapping their wings, like birds, and the hammerheads attacked them while they were swimming close to the bottom. Having selected its prey, the hammerhead dives steeply, butting the stingray with so much force that it strikes the bottom and bounces off again. But before the stingray can recover, the hammerhead delivers a second blow, this time following the action through and pinning it to the sea floor with its head. The attacker then pivots around to face its prey, still pinning it down, and seizes one of its wings in its jaws. Shaking its head violently from side to side, the hammerhead removes a crescent-shaped bite, releasing its prey. The shark then circles overhead, waiting. The stingray swims off, at a reduced speed, but is soon overtaken by the hammerhead, which attacks again, removing another bite. The second bite usually incapacitates the stingray, and the hammerhead can finish feeding upon it at its leisure.

The thresher shark is remarkable for its enormously elongated tail, which is longer than the rest of the body. The tail is used to thresh at the surface of the water, to frighten smaller fishes. It is said that threshers hunt in packs, using their tails to herd fishes, like mackerel, into a tight knot, which can then be attacked en masse. Unfortunately, we do not know very much about this unusual shark.

Killer whales (*Orcinus orca*) are about twice the length of white sharks; males, which are the largest, average about 27 feet (8.2 m) and weigh about 6 tons. Like the white shark, they include pinnipeds and whales in their diet, but their hunting techniques are radically different—killer whales are intelligent predators, though that does not imply they are any more successful. Their range extends from the tropics to the polar seas.

A pod of killer whales cruises off the rocky California coast on a spring morning, about one mile from shore. They are seven in number and include an adult male and two calves. He is easily distinguished from the females by his tall, straight dorsal fin. They are probably a family unit. Their course lies parallel to the shore, and they seem to be in no hurry.

They maintain the same course and speed for almost an hour. Then, for no apparent reason, they make a right-angled turn and head for shore. Their previous progress had been orderly, maintaining a discrete formation, but they are now spreading out on a broad crescentic front, leaping and gamboling in the water like children at play. And now the reason for their change in behavior becomes apparent: they have intercepted a large herd of sea lions. The sea lions are no match for the killer whales, and their only salvation lies in reaching the land. In their desperation to reach safety, they abandon their usual mode of swimming and are porpoising along at the surface at great speed. The whales swim beside them, dive beneath

Male (top) *and female killer whales.*

them, and cut in front of them, apparently having the time of their lives. The sea lions are obviously terrified, but the whales just seem to be playing with them. One of the whales leaps right out of the water, clean over the top of a fleeing pinniped, landing with a resounding splash several yards ahead of it. Sometimes a tormentor bumps up against a sea lion, knocking it off course. Another time a killer whale dives and surfaces directly beneath one, sending it flying into the air. These exuberant antics may appear to be random, but the killer whales' sport is not without purpose. Their crescentic formation has been getting narrower—they are herding the sea lions closer together.

The pinnipeds, now in a tight formation, are close to exhaustion, but their tormentors seem to have energy to spare. The game seems destined to continue until the sea lions have hauled up on land—they have only one hundred yards to go—but then the killing begins. Like so many cats with so many mice, the whales throw themselves at the sea lions, biting and slashing, mauling and dismembering, so that the sea foams red with their frenzy. Each whale spends the minimum amount of time with its prey before moving on to the next, and within a few minutes all the sea lions are dead.

Killer whales have curved conical teeth that intermesh, top with bottom sets, as the jaws are closed. The contact between opposing teeth is so tight that they rub against one another, grinding sharp wear facets along their edges. In contrast to the jaw joint of terrestrial carnivores, like lions, which permits only an up-and-down movement, that of the killer whales is more flexible, allowing the lower jaw to move from side to side as well as back and forth. This gives the jaws and teeth a powerful crushing and rendering action, evidenced by the considerable damage they can inflict.

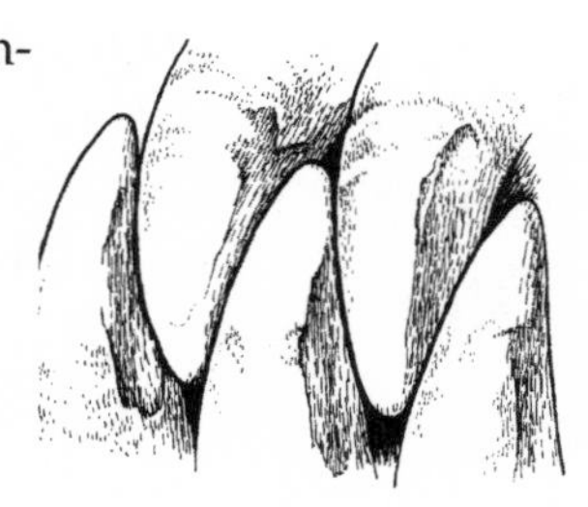

The teeth of the killer whale; note how they mesh together (top) *and how this grinds sharp wear facets along their edges.*

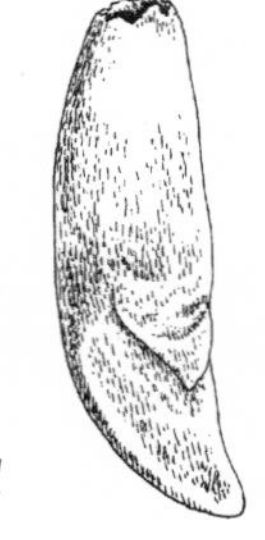

Torn bodies heave and tumble in the swell. A ragged torso, trailing a knot of bloated intestines, bobs against a severed head. Flocks of gulls swoop and dive, screeching raucously as they pick over the spoils. There is more than enough to satisfy their cravings, but they still squabble and fight over the meanest scraps of offal.

The whales, the killing done and their former demeanor restored, swim slowly through the carnage as if witnessing the deeds of others. They begin to feed, unhurriedly. Each one will leave with a full stomach.

Waves lap gently against the rocky shore. Crabs scurry, fishes dart, kelp fronds sway with the rhythm of the sea. Life goes on, oblivious of the carnage out in the bay. A starfish caresses a horse mussel in its pentagonal embrace. Its arms, anchored to the mussel by a myriad of tiny suckers, patiently pulls the two shells apart. A small opening has already appeared, and the starfish has started everting its stomach into the gap. Digestive enzymes from the disgorged organ seep into the mussel's soft interior, breaking down its tissues and weakening its resolve. Eventually the entire contents will be turned to liquid, which will be sucked up by the starfish's stomach. The starfish will then withdraw its stomach back through its mouth and glide slowly away. Death stalks the shore as inexorably as it does the sea. Only its tempo is different.

Our next encounter with killer whales takes us to Antarctica in November. It is late spring in the Southern Hemisphere, and the ice is beginning to break up.

A tall black fin cuts through the frigid waters, clearing a path through the scattered pebbles of floating ice. The sea is flat and calm, and the wake from the whale's passage ripples out on a broad front, causing the ice floes to bob majestically. The whale appears to be alone, but moments later four more whales surface about forty yards away. Their sleek black backs glisten in the sun with a metallic luster—they could be nuclear submarines. And just like some submarines, they have a dome up forward, housing sonar equipment. The four whales have surfaced in an open patch of sea and swim, unhurriedly, toward a flotilla of ice floes. The male dives, reappearing a few moments later beside the others.

One of the whales rises vertically out of the water, exposing the forepart of its body as far back as the forefins. This maneuver, called spy-hopping, is to enable it to look for prey above the surface. The whale scans a nearby ice floe, sees that it is barren, and moves on to the next one. Other individuals check out other ice floes in a similar fashion.

The pod has come to a halt beside a 20-foot-diameter (6 m) ice floe. Two of the whales are spy-hopping and can see a lone seal lying in the middle of the ice. One of the watchers moves in for a closer look. Its body rubs against the ice as it rises vertically in the water. The seal lies well above sea level, but the whale is so far out of the water that it stares down at its intended prey. The seal, safely out of the cetacean's reach, returns the whale's glassy stare.

The killer whales swim slowly around the floe, spy-hopping at random. The seal lies motionless, watching and breathing heavily. The whales' vigil continues for another five minutes, then abruptly ends as the whales dive and swim away. They resurface 100 yards (92 m) away, form up into a squadron in line with the ice floe, then dive again. They surface twice during their submarine dash for the floe, coming well out of the water each time. The last dive is made with great vigor, just 10 yards (9 m) short of the floating sanctuary. The result of their combined effort is a large wave that washes over the ice floe, tipping it up at such a steep angle that the seal is thrown into the sea. Whales and seal disappear beneath the surface, and the only evidence that anything has happened is the bobbing of the ice floes and the lapping of the waves.

One minute passes. Two minutes pass. Then the whales break surface two hundred yards from where the seal was last seen. One of the females has something in her mouth—a broken rag doll. With a flick of her head, she tosses the 500-pound (225 kg) seal carcass high into the air. She plays with the carcass for several minutes before it is ripped apart and eaten.

Considering their large size, killer whales are remarkably maneuverable. They can catch fishes, like salmon, and the larger species of penguin, which they swallow whole. But the smaller species, like the jackass penguin, can easily outmaneuver them. Killer whales can also perform some remarkable feats above the surface. They have been seen swimming after sitting ducks, successfully snapping them out of the air before the birds have gained sufficient height to escape. Even more surprising is their readiness to beach themselves to catch their prey. This strategy is used when hunting pinnipeds close to shore, either by single whales or by cooperative groups. The selected individual is chased toward the land, with the pursuing whale close behind. On reaching the shore, the whale is driven up the beach by its momentum, where it makes a grab for the prey. Since pinnipeds are rather sluggish in making the transition from swimming to walking, the whale stands a good chance of catching the seal in the surf. If other whales are participating in the chase, they stop short of the beach, stationing themselves in flanking positions to cut off any escape toward the sea. The beached whale lunges at its prey with a sideways motion of its head. If successful, it seizes the pinniped and holds it in its mouth until it regains the sea. Getting off the beach is achieved by flexing its back—head and tail raised—and rocking sideways to swing its body around parallel to the shore. The waves then help lift it off the shingle, and the whale swims clear. Killer whales are so adept at beaching themselves that they have been known to seize moose that are walking along the shore. But for all their predatory ways, they never seem to attack the human species. Why this should be so I have no idea. Perhaps it is because they are intelligent enough to recognize that we are not pinnipeds, even when we are dressed in neoprene wet suits.

Our last look at these remarkable predators takes us to the Mexican coast of Baja.

A heat haze shimmers across the white sand. Candelabra cactuses hold their prickling arms to the sky. And beyond the desert plants and the scorched beach lies the blue Pacific. The great ocean stretches for seven thousand miles, room enough for the largest animal that ever lived on Earth—the blue whale.

An adult blue whale can weigh over 200 tons, but this immature 60-footer (18 m) weighs closer to 45 tons. And it is being eaten alive. The attack, by over thirty killer whales, began two hours before. The beleaguered whale, trailing streams of blood from several wounds, is flanked on either side by three or four individuals. Two more swim ahead, and three behind. A squadron of five killer whales take turns patrolling under the blue whale's belly, preventing it from diving. Three more swim above its head, discouraging it from raising its blowhole above the surface, thereby hampering its breathing. Dominant males lead sorties to rip off slabs of blubber and flesh. They have already shredded its tail flukes, reducing its speed and making its swimming more laborious.

Three hours later: The blue whale has been under attack for over five hours and must be close to the end. Oblivious to the punishing attempts of two females to keep its head beneath the waves, the great beast thrusts its body skyward. Stale air explodes from its blowhole, sending up a great plume of spray. Much skin and blubber have been stripped from its head. A skein of loose skin from the females' last attack hangs like a tassel. Seawater cascades from its back, revealing a ragged wound where a dorsal fin used to be. The wound weeps blood in a steady stream. A flanking male works away at a gaping hole in the whale's side. Relays of killer whales have cut and thrust at the wound for over two hours, enlarging it until it is six feet across. The cavernous opening extends as deep as the muscle, pouring blood like an open tap.

Half an hour later: The blue whale has stopped swimming and is surrounded by its attackers. They feed unhurriedly, bolting down great gobbets of blubber and strips of skin. Some have dived beneath its head and are feeding on its tongue. The great whale is probably dead.

The slow demise of a great whale by a large pack of killer whales is reminiscent of the prolonged death of an African buffalo at the hands of a large pride of lions. The reason for the extended suffering in both cases is the large size difference between predator and prey. The weight of a killer whale is about one-tenth that of a blue whale. Even with their formidable teeth, it takes killer whales a considerable time to inflict sufficient damage to kill their prey.

I cannot leave the topic of killer whales without sharing a remarkable story from Australia. I must confess that when I began turning the pages of Tom Mead's book, *Killers of Eden,* I was skeptical. Here was an account of a pod of killer whales that returned to the same remote bay on Australia's south coast every winter, over a period of some eighty years. They returned to Twofold Bay, near the township of Eden in New South Wales, to hunt for large whales. But what is remarkable about the story is that the killer whales hunted in cooperation with the local whalers.

There was a shore-based whaling station at Eden from the mid-1800s to the early 1930s. Observers on land would scan the sea for whales. Once a whale was spotted, the whalers would set out to sea in large wooden rowing boats. The whales were harpooned by hand, and the whalers would let out enough rope to keep their boat well clear of the whale while it was towing them. At length the whale would tire, allowing the whalers to get close enough to deliver a mortal wound with a long lance. Sometime during the hunt the whalers would usually be joined by the killer whales, which would attack the doomed animal and try to drive it toward the shore. On other occasions the killer whales would be on the scene before the men, and if the whalers were unaware of the presence of a large whale, some of them would swim to the wharf and make a sufficient disturbance in the water to attract their attention!

Once the whale was dead, the whalers would let their cetacean helpers take their fill before attempting to tow the carcass back to the wharf. The killer whales would not eat very much—the tongue and perhaps some skin—a reasonable reward for all their work. The success of the whalers was largely attributed to the help they got from the killer whales. The last of the pod died in 1930, and the whaling station closed down two years later.

I am still not sure what to make of this remarkable story. The descriptions of the way the killer whales attacked the larger whales agree completely with later accounts. And we know from the experiences of trainers at marine aquaria how readily killer whales cooperate with humans. The story is certainly not beyond the bounds of credibility, and I recommend that you read the book for yourself and make up your own mind. [2]

Most of what we know about whales has been learned from studying dead ones—we have killed them in the hundreds of thousands, and mostly not from necessity. It follows that our knowledge of whales is far from complete, and perhaps the most mysterious whale of them all is the sperm whale (*Physeter catodon*). While not the biggest of whales—it reaches about 50 feet (15.3 m) in length and weighs about 40 tons—many of its other attributes can be described in superlatives. It has the largest cranium of any animal, living or extinct, and the largest brain (20 pounds, or 9.2 kg, compared with an elephant's 10 pounds and our 2 pounds); it is the largest toothed mammal; and it probably dives deeper than any other animal. During its deep dives, which last over one hour, it descends to depths of well over 3,280 feet (1,000 m).

Herman Melville's Moby-Dick was a sperm whale, which probably explains why its distinctive shape readily springs to mind when we think of whales. The great square head with narrow lower jaw that characterizes the species occupies about one-third of its entire body length. Imagine a

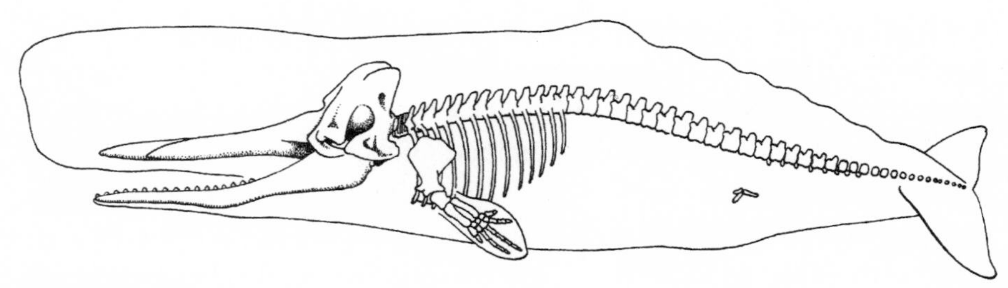

The sperm whale.

head almost two car lengths long and as high as a room. But why does a sperm whale have such a large head? Mysticetes, like the blue whale, need large heads because they need huge mouths for filtering large volumes of plankton. But the sperm whale is an odontocete; it does not filter its food and consequently has a relatively small mouth. It is true that the sperm whale has a large brain, but that occupies a very small part of the head. Most of the volume of the head is taken up by an oil-filled structure called the *spermaceti organ,* together with a second fatty mass that appears to be the equivalent of the porpoise's melon. These structures are associated with an intricate system of air ducts and spaces. The function of the entire complex is not well understood, but it appears to generate bursts of high-energy sonar sound. We will return to this apparatus later on but first need to consider the question of food and feeding.

Sperm whales have a predilection for squid, those torpedo-shaped relatives of the octopus that can propel themselves through the water by jet propulsion. Squid come in a wide size range, from small species that would fit in your hand, to 50-foot (15.3 m) monsters of the deep. Sperm whales eat them all. We know they tackle the giant ones because they sometimes bear the saucer-sized scars of their suckers on their bodies. Sperm whales also eat a remarkably wide variety of other seafood, such as tiny lantern fishes only an inch or two long, skates, puffer fish, mackerel, salmon, angler fish, tuna, barracuda, and sharks, including some large ones. This list encompasses animals from all levels of the sea, from surface-living species like the puffer fish, to animals of the deep like the angler fish. They also range in swimming speeds from slow-moving fishes like skates, which spend most of their time cruising slowly on the bottom, to fast swimmers like the mackerel and tuna. This wide array of prey species is all the more remarkable when account is taken of the sperm whale's limited food-gathering equipment. Its jaws are long, but very narrow, and stop well short of the front of the head. The teeth, which are well spaced, are confined to the lower jaw. They appear to play little or no role in prey capture because tooth marks are so seldom seen on the animals they have eaten. Indeed, it has been reported that live squid sometimes swim out of the stomachs of freshly harpooned sperm whales when they are cut open.

While squid spend much of their time cruising at a gentle pace, they are capable of remarkable bursts of speed, using their jet propulsion system, and have been reported to exceed 30 mph (49 km/h). Burst speeds of fast fishes like the tuna and barracuda are similarly impressive. The sperm whale, on the other hand, appears to be capable of burst speeds of only about 14 mph (22 km/h), spending most of its time cruising at a leisurely 2–3 mph (3.2–4.8 km/h). But speed is not the only attribute needed to catch fast-moving prey; a predator must also be maneuverable, especially when dealing with an agile swimmer like a squid or a mackerel. I have tried netting mackerel while they were swimming in a large holding tank, and can attest to the difficulty of the task. I might not be the fastest thing with a net, but I bet I am more agile than a 50-ton sperm whale! Given that a sperm whale's daily food requirements are about 1–2 tons, how can it possibly manage to satisfy its needs with such limited food-gathering abilities? That question brings us back to the spermaceti complex.

The primary role of the cetacean sonar system seems to be for echolocation. To detect an object by echolocation, the predator emits a sound, which bounces off the intended prey as an echo and returns to the predator. If all were quiet and you clapped your hands together near a large obstacle like a wall, you would hear an echo. But if you repeated the experiment in front of a smaller target, like a lamppost, you would probably be unable to detect an echo. The reason for this has to do with the wavelength, therefore the frequency, of the sound signal. High-frequency sounds, like the high notes from a violin, have a short wavelength, whereas low-frequency sounds, like the bass notes from a double bass, have a long wavelength. To produce an echo, an object has to be large enough to reflect the sound falling upon it, and that requires it to be at least as long as the wavelength of the sound. The frequency of a hand clap is fairly low; therefore the wavelength is fairly long. A wall is much wider than the wavelength of a hand clap, so it reflects the sound and produces an echo. But a lamppost is narrower than the wavelength, so it does not reflect the sound to produce an echo. The detection of small objects by echolocation therefore requires high-frequency sounds. But the downside of using high-frequency sounds is that they are readily absorbed by the surrounding medium, whether this be air or water, and therefore have a shorter range. The low-frequency

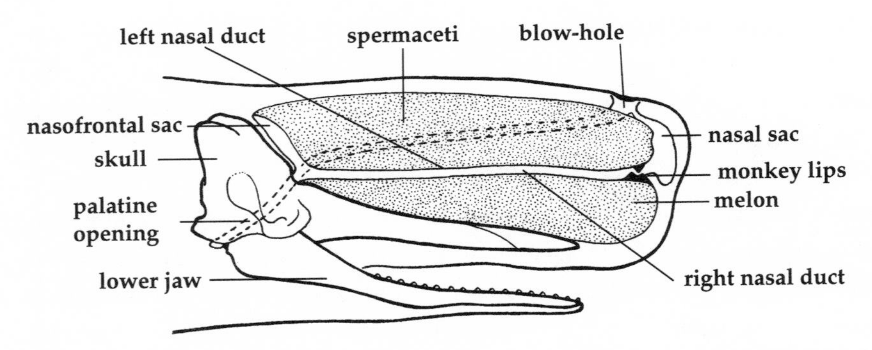

A diagrammatic representation of the spermaceti complex, and ducts, of a sperm whale.

trumpeting of an elephant or the throaty roar of a lion, for example, can be heard for several miles, whereas the high-pitched calls of monkeys and birds are soon lost.

We humans, like most other mammals, generate sound in the larynx, or voice box. Some researchers contend that odontocetes do the same, but this now seems unlikely. Although we are a long way from understanding the precise mechanism of sound production, a recent study, the culmination of a decade's work, shows that sound is probably produced within the forehead. As mentioned earlier, odontocetes have a complex system of nasal ducts and oil-filled structures on top of their skulls. The nasal ducts are paired, and toward the end of each one is a complex structure, named the *monkey lips* because of its appearance[3] (in the sperm whale there is only one set, associated with the right nasal duct). It is thought that air is forced down the nasal ducts, causing the monkey lips to vibrate, thereby emitting a burst of high-frequency sounds. These sound pulses are transmitted through the oil-filled structures to the surrounding water. This all takes place with the blowhole closed, and the air passing through the monkey lips complex gets recirculated.

Although a low-pitched hand clap is reflected from a distant wall, the echo appears to come from all around, and it is difficult to get a precise fix on its position. This is because the hand clap spreads out on a broad front. Pinpointing the direction requires a focused beam, and the narrower the beam the greater the precision. Beams can be focused using a reflector, as when we cup our hand to our mouth to shout. The reflector has to be at least as large as the wavelength of the sounds being focused, so low-frequency sounds require larger reflectors than high-frequency ones. An animal that echolocates therefore needs to generate a narrow beam of sound, and to use those frequencies that give the optimum range for the size of the targets to be detected. (Incidentally, since sound travels faster in water than in air, a given frequency sound has a shorter wavelength in water than it does in air.)

For underwater sonar, frequencies need to be below 10 kHz (10,000 cycles per second) to give a sufficiently long range. The sonar clicks of sperm whales appear to be of this order, peaking in the 1–5 kHz range. The wavelengths of these frequencies range from 1 foot to 4.5 feet (0.35–1.4 m) and would therefore allow the sperm whale to detect fairly small targets. On the other hand, the low end of this frequency scale would require a large reflector. The spermaceti complex appears to be involved in focusing the beam, which is probably one of the reasons for its large size. The spermaceti complex also generates the sound, and one of its obvious features is a massive muscle layer that stretches from the top of the skull to the front of the spermaceti organ. It has been estimated that this muscle could jerk the spermaceti organ backward with a force of 10 tons, producing such high pressures in the air spaces as to generate a high-intensity sound beam. But is there any evidence that sperm whales do generate loud sounds?

A few years ago a juvenile sperm whale got itself trapped in a New York yacht basin. This was inconvenient for the whale, but it did allow a rare opportunity to investigate a living sperm whale. The series of clicks that the whale made were so loud that they could be heard across the entire area. Investigators tried to find out where the sounds were coming from by running their hands over the animal's head. They found that the highest intensities were located in a circular region over the forehead; the force was so great that their hands were knocked away from the animal's head.

Sound intensities are measures in decibels (dB). Normal speech has an intensity of about 50 dB, a rock concert reaches about 120 dB, while a jet engine peaks at about 160 dB, which is beyond the threshold of pain. In captivity, odontocetes generate clicks in the range of 140–180 dB, which is very loud, and similar high values have been recorded in the wild. In experiments where dolphins have been trained to locate small targets against high background noises, they can generate clicks with intensities as high as 228 dB. Calculations have shown that sperm whales should be able to generate even louder sounds, with intensities as high as 265 dB. What is particularly interesting here is that such high-intensity sounds can be lethal. An effective, but unsporting, way of fishing is to drop explosives into the water. If explosives like TNT are used, the fishes are killed by sound intensities of about 230 dB. This level is comparable to the loudest sounds dolphins can make, and less than the estimated sound intensities of sperm whales. Could sperm whales, and other odontocetes for that matter, use sonar for stunning their prey? Is that the significance of the sperm whale's huge head? If sperm whales did use their sonar for incapacitating their prey, it would explain the remarkable size range of the animals they eat. It would also explain why they do not appear to use their teeth on their prey. While the idea of animals stunning their prey with a sonic beam seems fanciful, I should point out that it is not without precedent. There is a shrimp, for example, called the pistol shrimp (*Alphaeus californiensis*), that produces sharp cracking sounds underwater by snapping its enlarged pincers. These sounds are so intense that they can stun small fishes, a phenomenon that can be demonstrated by putting a fish into an aquarium containing a pistol shrimp. The shrimp crawls toward the fish with its pincers extended, and, when close enough, it snaps them shut. The shock wave stuns the fish, which the shrimp then starts eating.

Experiments on the stunning abilities of captive dolphins have so far been inconclusive. In one such experiment a dolphin was allowed to swim in a tank with a school of fishes. The fishes were not stunned by the dolphin's sonar clicks, but some of them became so disoriented after two hours of trials that they could no longer stay within the school. The fishes that left the protection of the school were soon snapped up by the dolphin. There have also been reports of fishes becoming lethargic after dolphins had passed through their schools. In one instance the fishes were so sluggish that they could be caught by hand. There is also an account of a salmon that stopped swimming, for no apparent reason, after being followed by killer whales. The salmon, which lay motionless in the water, was snapped up by one of the killer whales moments later. This last observation helps explain how a 9-ton killer whale can catch a considerably smaller and more maneuverable swimmer like a salmon.

Imagine a sperm whale making one of its long dives. It might be swimming in a blue tropical ocean or in the cold green Atlantic. The great square head is lowered beneath the waves, and the last thing seen from above the surface is a huge triangular tail sticking straight out of the water. For the first 100 feet (31 m), the water is aquamarine, gradually changing to a rich cobalt blue. And as the sea changes to a deeper shade of blue, the temperature falls. Dark blue gives way to blue-black, as the whale approaches the realm of perpetual night. Its eyes are of little use now, and it relies on its sonar to direct its path. Brilliant flashes of light stab out of the darkness—

they are as varied and fantastic as the unseen creatures that wear them. And what brings the sperm whale to this cold world of perpetual night? Perhaps to hunt for giant squid, but we cannot be sure. Like so many other things in the ocean, the ways of the sperm whale remain an enigma to us.

One mystery of the sea now appears to be close to a solution. Whatever creatures are responsible for mutilating the seals of Sable Island, they are obviously not white sharks. But Zoe Lucas became convinced that they are some kind of shark, and tried to find out which one. To this end, she contacted shark researchers in South Africa, Australia, and the United States, sending them color photographs of the mutilated seals. She also consulted with the medical examiner for Dade County, Florida, a state that has extensive records of shark-attack victims on file. Like a homicide detective tracking down a serial killer, she followed every conceivable lead, but most of them led down blind alleys. And then the trail took a turn to the far north. She contacted Norwegian fishing captains who were working in Davis Strait, the seaway between Greenland and the Canadian Arctic. Lucas had a suspect: *Somniosus microcephalus*, the Greenland shark.

The Greenland shark, a close relative of the common dogfish, is a cold-water species of arctic and subarctic regions which rarely strays far south. Said to be sluggish, it is often referred to as the sleeper shark and appears to spend much of its time resting on the bottom. The average length is about 10 feet (3 m), but much larger specimens exist: there are many

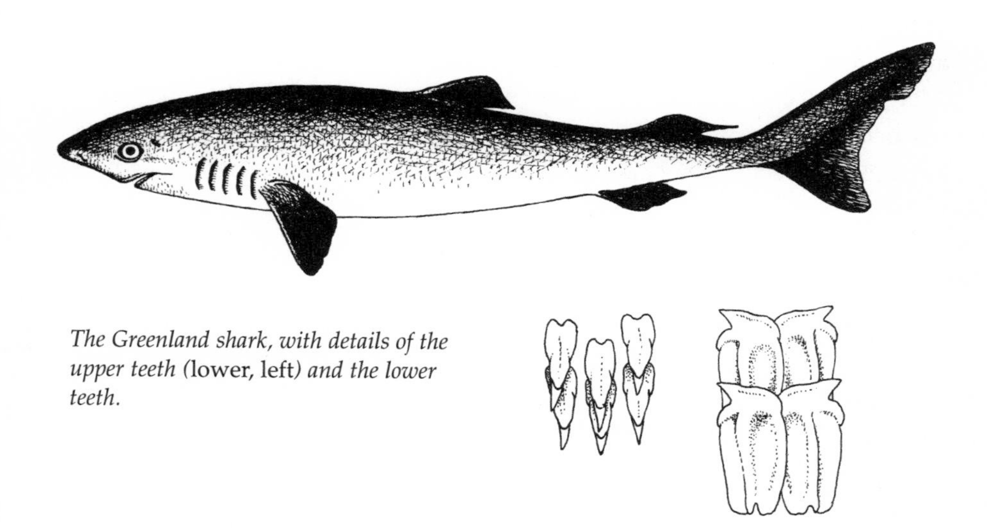

The Greenland shark, with details of the upper teeth (lower, left) *and the lower teeth.*

records of individuals in the 16- to 18-foot range (5–5.5 m), and the largest recorded was 21 feet (6.4 m). Greenlanders, setting out to sea in their kayaks, have fished this shark for over a century, primarily for its liver. The animal is so sluggish that it can apparently be hauled alongside a kayak, slit open to remove the liver, and released again, without mishap to the fishermen. In spite of its evident sluggishness, it is a predator, feeding on a wide variety of prey, including cod, haddock, halibut, skate, various shellfish, crustaceans, seabirds, and seals.

The teeth of the lower jaws are oblong, with short cusps pointing outward, forming a cutting edge like a serrated knife. Lucas conducted some experiments with isolated jaws and showed that the lower teeth could cut through seal hide and blubber. The upper teeth are slender and pointed, and although they might be able to puncture the hide, she thinks it unlikely they would be able to cut through it. How could a Greenland shark inflict the spiral wounds? Lucas thinks the shark would engulf the seal in its mouth, perhaps while lying in ambush on the seabed. The shark would then sink its lower teeth into the blubber of the seal's torso. Then, by a powerful twisting action of its own body, the shark would inflict the spiral cut in the blubber, thereby peeling the upper half of the seal's body, like an orange. In this way the seal would be partially stripped of its energy-rich blubber, which the shark would devour. The rest of the carcass would be discarded, to be consumed by scavengers or washed up on the beaches of Sable Island.

Some sharks often leave shed teeth, or fragments of broken teeth, in their victims, and these can then be used to identify the species. But no such weapons have been found on the bodies of the mutilated seals. Nor are there any official records of Greenland sharks having been caught off Sable Island, but recent reports from fishermen indicate that they may have been. The case against the Greenland shark is only circumstantial at present, but I suspect it is only a matter of time before incriminating evidence is found, and the case can be closed.

4

Avoiding Attention

I grew up with camouflage. The roof of our local hospital was painted green and brown, a mottled livery it wore until long after the end of World War II. Similar color schemes were used to camouflage military equipment, from Jeeps and tanks to ships and planes. The idea of camouflage is simple enough: the object to be hidden is made to blend in with its surroundings. It is not difficult to visualize how the mottled hospital roof would blend in with the surrounding parkland when viewed from the air, just as a caterpillar blends in with its foliage. But chances are that it did not! Hugh Cott, a zoologist at Cambridge University, made an extensive study of animal camouflage, publishing a book on the subject in 1940. This was during the early days of the war, and he was very critical of the attempts that had been made at camouflaging buildings and equipment. His main point was that our version of camouflage, the familiar blotches of green and brown, was not sufficiently bold. What was needed, instead, was a disruptive color pattern, as used by many kinds of animals. Such striking patterns, like the bold stripes of tropical fishes and the contrasting color patterns in birds, look glaringly conspicuous when viewed from up close. But they are meant to be seen from a distance. The bold pattern then breaks up the animal's outline, making it melt into the background.

Disruptive coloration, in combination with a few other basic strategies, can result in some astonishing acts of disappearance within the animal

Bold stripes, as in the killdeer and the zebra butterfly, help break up an animal's outline when seen from a distance against a variegated background.

kingdom. But not all animals are camouflaged. Some, like the wasp and the peacock, are conspicuously colored to advertise their presence. In the wasp's case this is to warn would-be predators that it is dangerous, while the peacock's point is to show potential mates how superior he is to the other males. At the other end of the spectrum to these extroverts are animals like stick insects, which are so perfectly camouflaged that their bodies are shaped to look like part of their environment. Regardless of these extremes, though, most animals are camouflaged to a greater or lesser extent, either to help them hunt without being detected or to avoid being preyed upon themselves. This is quite apparent from a walk in the woods, where you are likely to hear more animals than you ever see.

Some animals, like stick insects, have bodies that are shaped to look like part of their environment.

Most of the examples of camouflage seen in nature concern visual perception. This requires that the animals that are to be deceived have good eyesight, but this is not always the case. Many marine invertebrates, for example, have rudimentary eyes. A starfish's eyes are only capable of distinguishing between light and dark, and they locate their prey by scent. It would therefore be of no benefit to the molluscs it hunts if they were camouflaged. But if they were able to release a false scent into the water, this would help them to confuse the predator. And this is precisely what some of them appear to do, as we will see later on. To take an example from our mechanized world, World War II pilots attempted to make visual contact with enemy aircraft, and airplanes were therefore visually camouflaged, to reduce the chances of being detected. Modern aircraft, in contrast, are detected by radar, so stealth technology was developed to make aircraft invisible to radar, rather than to human eyes.

The luminous images that Joseph Turner captured on canvas during the last century arose from his masterly control of light. When light falls on a three-dimensional object, it casts shadows. An apple lying on a table casts a shadow not only on the tabletop but also on its lower half, so that the top of the apple appears lighter than the bottom. Shadows emphasize the three-dimensional nature of a solid object, and if these can be eliminated the object begins to lose its solidity. The gradation in light from the top to the bottom of an object can be eliminated by *countershading,* in which the upper segment, which receives most of the light, is shaded darker than the lower segment. Countershading is so common among animals that we have probably all seen some examples. Many land mammals, including deer and hares, are darker on their backs than on their flanks and undersides. Similarly, many birds, including sparrows and gulls, are lighter on the breast and the underside of the wings and tail than they are on their backs. Countershading is also common among pelagic fishes, where gray, black, or blue backs shade into white or silver bellies, as in herring and sharks.

However, countershading in pelagic fishes, and in seabirds, probably has not evolved for eliminating shadows, as when they are viewed from the side, but rather for blending them with the background when they are seen from above or below. Viewed from below, the white ventral surface of a shark blends in with the light filtering down from the surface. Conversely, when

Countershading is common among a wide range of animals, including the black-headed gull and the blue shark.

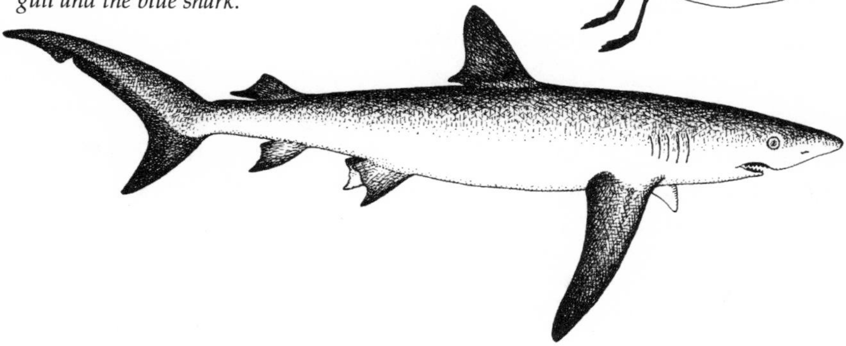

viewed from above, the dark dorsal surface blends in with the darkness of the deep. The same principle holds for many seabirds and has been applied to camouflaging aircraft, the underside being painted to match the sky, while the upper surfaces blend in with the ground.

The effectiveness of countershading in concealing a seabird's approach to its prey has been demonstrated experimentally. In this experiment, conducted on black-headed gulls (*Larus ridibundus*), half of the birds had their white underparts dyed black, while the others were left as they were. The hunting success of the two groups was then compared by letting them loose, one group at a time, in an enclosure housing a large and well-stocked fish tank. The normal birds consistently outperformed the dyed ones because they were less conspicuous. The fishes therefore had less warning of the attack and were less successful in escaping. The results help explain why many seabirds have white underparts.

I want to discuss countershading a little further, but before doing so I want to take a short diversion to explain a remarkable special case, called *counterillumination*, which involves *bioluminescence*. Bioluminescence, the production of light by living organisms, is exemplified on land by fireflies, but the most dazzling display of this phenomenon occurs at sea. I have sailed beneath the stars and watched the tropical sea explode in a blaze of phosphorescence as I towed a plankton net behind the stern. And when the glowing net was swung aboard, the light from a myriad of microscopic organisms shone so brightly that a book could have been read by their light. Bioluminescence occurs among a wide variety of marine animals, including many deep-sea fishes, some of which appear to use their ghostly light to countershade their bodies. Space limitations do not allow anything but a brief account of these remarkable fishes, but I will begin with a little more about bioluminescence.

The chemistry of bioluminescence is poorly understood but appears to involve two compounds, *luciferin* and *luciferase*, which react together to emit light. There also appear to be stable complexes of luciferin and luciferase that can be triggered to give off light by other substances, such as calcium ions.[1] In bygone days it was feared that the intense sparks of light could cause fires, but it was subsequently realized that the light is not accompanied by any heat.

Bioluminescent animals restrict their light displays to specific parts of their bodies, called *light organs*, or *photophores*. The hatchet fish (*Argyropelecus aculeatus*), a small, deepwater form, has numerous photophores, mostly arranged along the bottom of its body and pointing downward, but with one pointing into each eye. The fish can regulate the amount of light emitted from its photophores, matching this very precisely to the light levels filtering down from the surface above. It also matches the color of the light, and the angle at which it rakes down from above.[2] The photophore in front of each eye may play a role in this, allowing comparisons between ambient levels and its own light output. The fish is therefore able to counterilluminate its body, making it less conspicuous against the daylight when viewed from below. The hatchet fish feeds mostly on midwater invertebrates, and its photophores are almost certainly an antipredator system, rather than a device for allowing it to steal up on its prey.

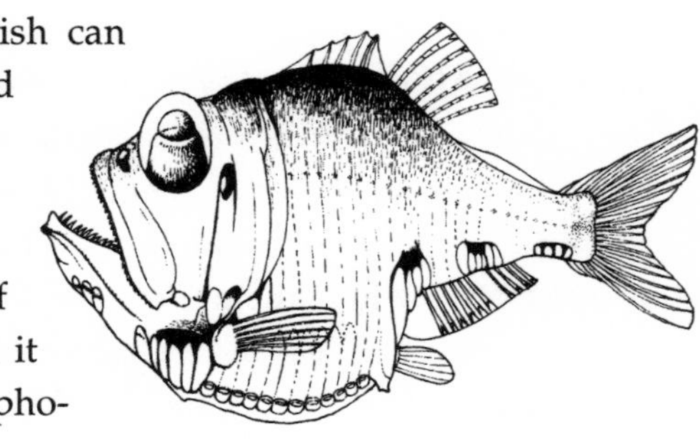

Most of the hatchet fish's photophores are arranged along the bottom of its body, but there is also one in front of each eye.

The hatchet fish's photophores are powered by its own luciferin-luciferase system and are therefore under the fish's own control. Such photophores, referred to as intrinsic, are found in many other fishes, and also in a wide variety of other animals including squid, jellyfishes, crustaceans, sea squirts, and worms. There is also a second type of photophore, powered by bioluminescent bacteria that live in a symbiotic (mutually beneficial) relationship with the animal that houses them. Bacterial photophores appear to be restricted to a few species of squid and fishes. There are usually only one or two photophores per individual, and they generate light constantly. Since the bacteria glow continuously and are not under the direct control of the animal possessing them, their light output has to be regulated. This is achieved by covering the light organ with shutters, or with *chromatophores* in the overlying skin, or by some other means. Chromatophores are amoebalike cells, lying deep in the skin, that contain pigment granules.

One particular deep-sea species (*Opisthoproctus soleatus*), sometimes known as the spookfish, possesses some of the most remarkable features seen in any fish. It has a large pair of tubular eyes, permanently set to gaze toward the surface, and a single bacterial photophore, located near the base of its tail. The bottom of the fish is flattened into a thin plate, called the sole, which extends forward almost as far as the level of the tip of the snout. Light from the photophore is channeled through an elaborate reflective tube to illuminate the entire surface of the sole. This bizarre arrangement functions like the multiple photophores of the hatchet fish, counterilluminating the ventral surface of the fish, thereby reducing its silhouette, when viewed from below, against the light from above. This same principle of counterillumination was proposed for use on aircraft during World War II, where the underside of the wings were to be illuminated by lights. The system was tested on torpedo bombers, used for attacking submarines on the surface. Although the system worked, it was made redundant by radar equipment before it went into service.

Countershading to obliterate shadows is beautifully exemplified by the hawkmoth caterpillar. These plump caterpillars feed on the leaves of several types of trees, including willows and poplars, and occur in various shades of green. They usually hang upside down from the foliage when they are feeding and are therefore reversely countershaded, being lightest dorsally, shading into a darker tone ventrally. When they are upside down, the countershading eliminates the shadow on their back, giving them more of a two-dimensional appearance, like the leaves around them. Their resemblance to the leaves is enhanced by the oblique stripes running along their sides, and by the color match of their bodies. Color matching is partly brought about through the caterpillar's diet. Certain pigments occurring in the trees are extracted from the leaves, to varying degrees, and deposited in the caterpillar's skin. Field observations show that caterpillars not closely matched to the color of the foliage are more readily preyed upon by birds than those that blend in well with the leaves—natural selection in action.

Color blending is such a general principle of camouflage that the natural world is replete with examples. Snakes and lizards that live in the desert, for example, tend toward earth tones—browns and grays and yellows—whereas those of the for-

The caterpillar of the hawkmoth spends most of its time upside down, its body reversely countershaded, making it least conspicuous when in this position (top).

The arctic fox in its winter coat of white is difficult to see against the snow.

est are brightly colored, predominating in green. I was reminded of just how vibrant these colors appear during a recent trip to the zoo, when I visited the reptile house. I saw a tree python there that was such a vivid shade of green that it looked quite unnatural, especially within a terrarium. But if it were viewed against a lush tropical forest, it would melt into the background. Bright greens are also worn by many other animals of the tropical forest, including tree frogs, birds, and insects. At the Earth's poles, animals tend to be white. Polar bears are permanently white, but many other animals living in the far North are white only during the winter months. These include the arctic fox, arctic hare, weasel, ermine, and ptarmigan. The pups of several species of seals are also white, making them less obvious to prowling polar bears, which are themselves inconspicuous against the ice and snow. But the young seals change their diaper whites for business-suit grays when they abandon the ice for the adult world of the sea.

The big cats occupy a diverse range of habitats, and this is reflected in the differences in their coat colors. The lion, an inhabitant of dry open grasslands, has a tawny pelage, which blends in well with the dry vegetation and muted earth tones. The leopard, in contrast, spends much of its time in trees and has a spotted coat to match the foliage. And the tiger's stripes blend in well with the tall grass in which it lives.

Some animals, mostly aquatic ones, have transparent, or partly transparent, bodies, which make them very difficult to see. They include many jellyfishes, sea squirts, shrimps,[3] some worms, and a few fishes. The caterpillar of the angle shades moth (*Phlogophora meticulosa*) has a transparent body, which makes the greenery in its gut visible from the outside. Since the gut occupies most of the body volume, the caterpillar is perfectly matched to the leaves upon which it feeds. This may be the simplest case of color-matching the body to the environment, but the most familiar example is probably that of the chameleon.

Chameleons, of which there are many different species, are lizards that are adapted to living in trees, where they feed on insects. Their fingers and toes are opposable, giving them a good grasp on the branches, and their tails are prehensile, too. They move very slowly and deliberately, freezing between advances, their hunting strategy being to approach their insect prey as closely as possible. The chameleon catches its prey using its tongue, a highly elastic structure that can be flicked out farther than the length of its body.[4] The tongue reaches a top speed of around 13 mph (21 km/h), and it achieves this in an interval of only about 20 milliseconds, at a staggering acceleration of 1,583 feet per second per second (486 meters per second per second). If it could maintain this acceleration for a whole second, the tongue would exceed a speed of 1,000 mph (1,750 km/h). The tongue's range is fairly short, only about 1 foot (35 centimeters [cm]), which explains the chameleon's stealthy approach and need for camouflage. Its ability to change color is shared with a wide array of other animals, both vertebrate and invertebrate—from cephalopods (octopus, squid, and cuttlefish) to fishes, amphibians, and lizards—and involves the same basic mechanism of chromatophores. The granules within a particular chromatophore are all the same color and can be concentrated within a small clump, reducing the intensity of that particular color, or spread out throughout the cell, increasing the intensity. In some animals, including chameleons and fishes, the chromatophores are under the direct control of the nervous system and can respond very rapidly, producing color changes within seconds. The chameleon's remarkable ability to change color is primarily used for offensive purposes, enabling it to blend into its surroundings so that it can steal up on its prey, but it can also be used defensively. When startled or threatened, chameleons undergo a dramatic color change, becoming almost black. They also puff themselves up with air to almost twice their normal size, open their mouths widely to reveal a brightly colored interior, and hiss menacingly. The entire display is a bluff because they have no means of fighting back, but the overall effect can be very intimidating and often has the desired effect of frightening off a would-be attacker.

I teach a field course on Canada's east coast most summers, and my students and I sometimes catch the occasional squid in the fish trawl. Squid, and other cephalopods, have chromatophores that are surrounded

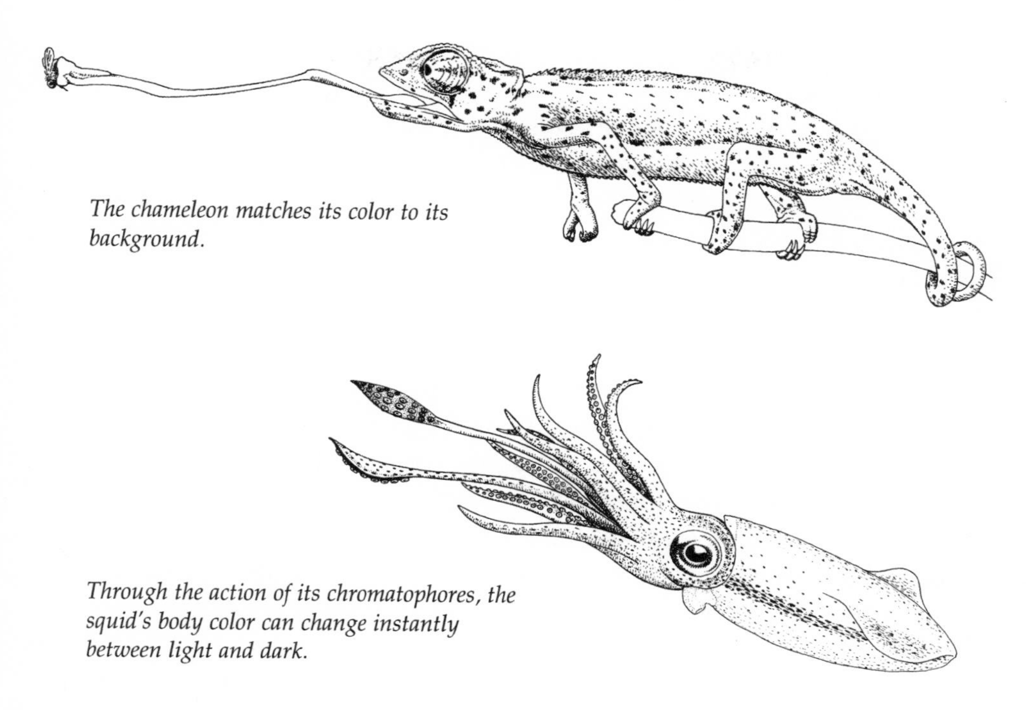

The chameleon matches its color to its background.

Through the action of its chromatophores, the squid's body color can change instantly between light and dark.

by small muscle fibers, and contraction of these fibers causes an instant change in color, which spreads across the body in waves. Their background color is white, mottled golden brown, and when they are handled, waves of color instantly flash across the body. One year we managed to rear some eggs, which hatched out as perfect miniatures of the adults, only a few millimeters long. Their chromatophores were the same size as the adults', and, since their bodies were so small, they had only a few of them. They were fully functional, though, and it was a great joy to watch them under a low-powered microscope, instantly dilating and contracting as the diminutive squid blushed and blanched. Squid are active predators, feeding on fishes and other fast swimmers. They seize their prey using a pair of extra long tentacles, killing and tearing them apart with their beaks. Their camouflage, like that of the chameleon's, is used both offensively and defensively, making them more difficult to be seen by potential prey and predator alike. They also have a second defensive strategy when attacked, discharging a black cloud into the sea from their ink gland. This cloud hides them from view while they make good their escape using jet propulsion. Their close relative the octopus is a sit-and-wait predator. These reclusive animals lurk on the seabed, waiting for suitable prey to come within grasping distance of their tentacles. They are masters of disguise, able not only to change color to suit their surroundings, but also to change the texture of their skin.

We have seen how countershading and color matching can help animals disappear into the background, and now need to consider disruptive coloration. Hugh Cott's criticism of wartime camouflage, with its muted browns and greens, was that it did not break up the outline of the object that was supposed to be hidden. Good camouflage, like good conjuring, hinges on delusion. The eye has to be tricked into not recognizing what it is seeing. This can be achieved by breaking up an object's outline with some unexpected features, features that can be found in the general background. Our local hospital stood beside a park, with a pond, and was close to several roads. Painted in typical camouflage colors, the rectangular roof would have looked like a mottled green rectangle when viewed from the air. But add a bold dark line, to simulate a road, and a silver blob for a pond, perhaps with a bright red square for an English telephone box, and a German pilot's eyes might just have been deceived. The workmen applying the paint would have questioned the sanity of their orders, and the roof would certainly have looked strikingly absurd from their perspective. But the whole point of disruptive coloration is that it *does* look glaringly obvious at close quarters but is convincing from afar. The conspicuousness of disruptively colored animals when seen at close range may be important for communications within species. The male pied flycatcher (*Ficedula hypoleuca*), for example, like many other birds, is more conspicuously colored than the female, and this coloration has probably evolved to attract mates. However, its conspicuous black-and-white pattern is disruptive, and the male is no easier to see when viewed from a distance than his more drab partner.

For disruptive coloration to be successful, the general color of the animal must blend in with its background, and the disruptive markings should be used sparingly and contrast maximally with the general body color. Thomson's gazelle, for example, is a light cinnamon brown, to match the earth and dried grass tones of the savanna, with prominent dark stripes on the sides,

Thomson's gazelle has prominent dark stripes on its body.

These drawings by Hugh Cott show how the conspicuous stripes of Thomson's gazelle can break up its outline when viewed from a distance against its natural background.

legs, and face. Seen from up close, the discordant stripes make the antelope stand out prominently. But when viewed from a distance, the stripes, acting in concert with dark objects in the background, have a disruptive effect, serving to break up the animal's outline. Cott gave some simple diagrams to illustrate his point, which are worth reproducing. In the first diagram he depicts a light-colored antelope against a featureless white background. In the next diagram he adds some dark disruptive stripes, which make the animal look even more prominent. However, when some dark elements are added to the background in the last diagram, the antelope's outline breaks up, making it much more difficult to see. If the antelope were from a forest rather than an open grassland, its body color would be much darker, to blend in with its darker background. The disruptive stripes would therefore be light, to give them maximum contrast. The bongo (*Tragelaphus euryceros*), for example, an antelope that lives in the forest, is a rich chestnut brown, with light disruptive stripes.

The bongo, which lives in the forest, has a dark coat with light, disruptive stripes.

Disruptive bars and stripes are so common in nature that we have all seen numerous examples: butterflies, tropical fishes, frogs, grasshoppers, birds, and zebras. Few animals on the African plain look more conspicuous than a zebra, with its prominent black and white stripes, and this is true whether the zebra is seen from up close or from afar. This has caused some biologists to question whether the zebra's stripes have got anything to do

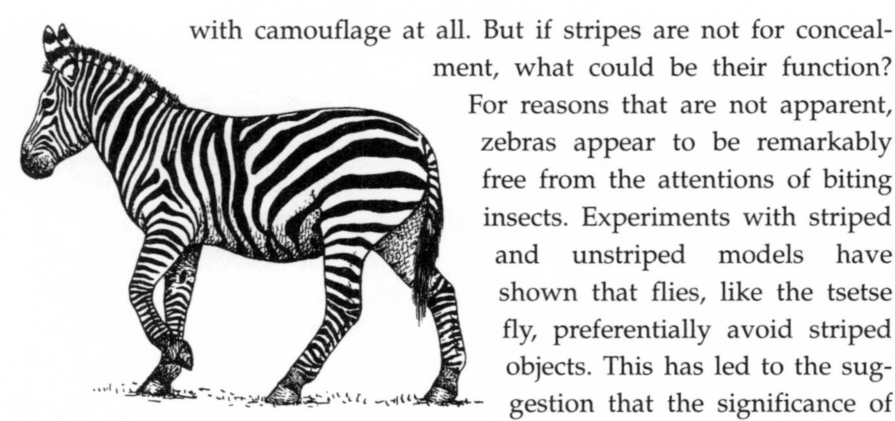

The zebra's stripes may serve more than a camouflage role.

with camouflage at all. But if stripes are not for concealment, what could be their function? For reasons that are not apparent, zebras appear to be remarkably free from the attentions of biting insects. Experiments with striped and unstriped models have shown that flies, like the tsetse fly, preferentially avoid striped objects. This has led to the suggestion that the significance of the zebra's stripes is to avoid biting insects, rather than large predators. Regardless of the effectiveness of the zebra's stripes in repelling biting insects, it seems that they do indeed have a concealment function. However, this is only apparent during times of low light levels, when most predatory attacks on zebra take place (mostly by lions). Under these conditions, zebras are said to be very difficult to see.

> "The stripes of white and black so confuse him with the cover," wrote one hunter, who had claimed to have seen thousands of zebra, "that he is absolutely unseen at the most absurd ranges. Time after time not only did Cuninghame and myself fail to make them out even as close as forty and fifty yards, but Kongoni confessed himself baffled. And of the many keen-eyed savages with whom I have had to do, Kongoni can see farthest and best."[5]

Disruptive coloration is often cited as a key factor in animal camouflage. However, I would like to add a cautionary note, precipitated by a field experiment on some butterflies in the tropics. These butterflies (*Anartia fatima*) are darkly colored, with prominent light, and presumably disruptive, bars on their wings. In the experiment, carried out in Panama over a period of five months, live butterflies were captured and released again after being painted with a black felt-tipped pen. Half of the butterflies had their light bars blacked out, while the others had the pen applied only to the dark areas, next to the light bars. The only difference between the two groups of butterflies was that one of them had their light bars obliterated. Since both groups had ink on their wings, they could be recognized again when they were recaptured. The experimenters expected that the butterflies with the blacked-out bars would have been less well camouflaged than the others and would therefore have been more readily eaten by the birds. The recaptured butterflies should then have included a higher percentage of the ones with the intact light bars. They were therefore surprised to find similar numbers of both kinds. The experiment seems to show that the disruptive bars made no difference to the effectiveness of the butterflies' camouflage. The experimenters were reluctant to cast doubts on the concept of disruptive coloration on the evidence of a single experiment, but thought it important to point out the need for caution in interpreting animal coloration. One possible explanation for the unexpected results is that the birds may have been avoiding the black forms because they looked so different from the others. We saw in chapter 1 that a wildebeest whose horns had been painted white was apparently readily picked out from the pack by predators. However, being conspicuous does not necessarily draw the attention of predators, especially for prey species that are solitary rather than communal. In an experiment on European blackbirds, for example, the wings of some birds were painted with a bright patch of color. Although this made them very conspicuous, hawks tended to avoid attacking them, presumably because they looked so different from their normal blackbird fare.

The eye, being a conspicuous and regularly shaped feature of an animal's head, needs special attention to reduce its prominence. This objective is usually achieved by the use of a prominent stripe, as we have seen in the illustrations of Thomson's gazelle. Some animals, especially fishes, have false eyespots, which may serve to draw attention away from the true eye or head. They may also startle the predator. Butterfly fish, for example, frequently have false eyespots. These reef fishes—often seen as colorful additions to marine aquaria—are deep bodied and have erectile dorsal spines, making them difficult and potentially dangerous to swallow. When threatened, they turn sideways to their aggressor and raise their spines. The false eyespot, which may be permanent or which may appear only at night, probably makes them look much larger than they really are by

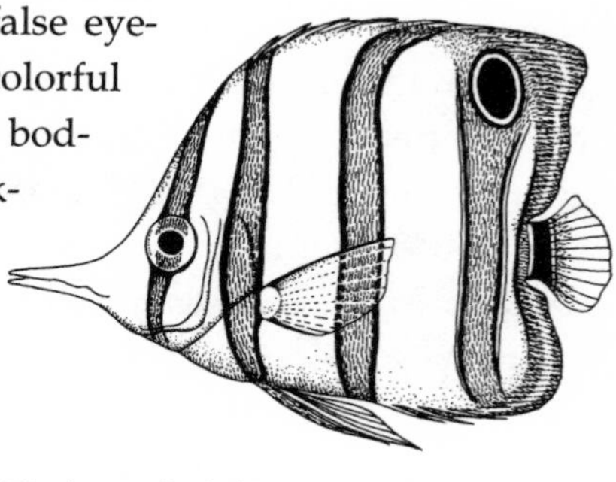

The butterfly fish's eye is partly concealed by a prominent stripe; a false eyespot draws attention away from this vital area of the body.

making their entire body appear to be just the head. The eyespot may also function as a warning signal, reminding the predator that these fishes are not good to eat. And if all this fails and the predator does attack, it is likely to bite at what appears to be a vital area, namely the eyespot (the true eye is usually camouflaged by a disruptive stripe). The fact that butterfly fish are often found with a segment missing from the body where the eyespot used to be is evidence of the effectiveness of this strategy. An example of false eyespots being used to startle a potential predator is provided by the mantis *Pseudocreobotra wahlbergi*. This winged insect, whose appearance is as terrifying as its Latin name, has large false eyespots on its wings. It threatens its aggressors by facing them with raised wings.

Animals that spend all of their lives against one particular background can be permanently blended to their surroundings, like the sandy-colored lizards that dwell in the desert, or the bright green tree snakes that live among the leaves of the forest canopy. It is under these conditions that camouflage reaches the heights of perfection and an animal can disappear from view like a conjurer's rabbit. Examples abound, from disruptively colored moths on tree bark to bitterns among the reeds. Bitterns, like their close relatives the herons, are adapted to a life beside the water, their long legs and necks, slender bodies, and sharply pointed bills being adapted for catching fish. They are equally adept at slipping through the reeds and are disruptively colored to blend into their surroundings. Viewed from the back, they are predominantly brown, but the front of the body is pale, disruptively marked by broken vertical bars. Part of their concealment strategy is behavioral, and when disturbed they shoot their bill skyward and freeze. The disruptive coloration of the long neck and breast blends into the reeds, making them virtually impossible to see. Since their fronts are better camouflaged than their backs, they turn toward the source of danger, and if that changes position, they turn too. The lengths to which these cryptic birds will go to conceal their presence is well illustrated by the observations of an American biologist, W. B. Barrows, writing in 1913. Barrows, in the company of another, had just watched an American bittern (*Botaurus lentiginosus*) land on a lily pond and adopt the characteristic freezing posture on being approached:

> . . . as we stood admiring the bird and his sublime confidence in his invisibility, a light breeze ruffled the surface of the previously calm water and set the cattail flags rustling and nodding as it passed. Instantly the Bittern began to sway gently from side to side with an

> undulating motion which was most pronounced in the neck but was participated in by the body and even the legs. So obvious was the motion that it was impossible to overlook it, yet when the breeze subsided and the flags became motionless the bird stood as rigid as before and left us wondering whether after all our eyes might not have deceived us.

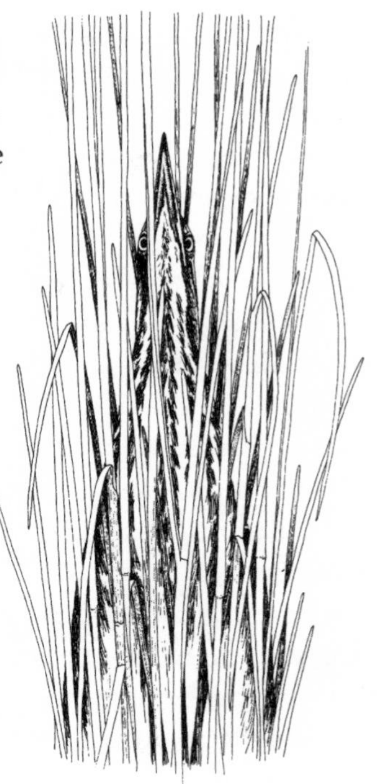
The bittern's disruptive coloration and vertical stance allow it to blend into the reeds.

The thought crossed their minds that this might have been an illusion, caused by the shimmering motion of the reeds and the rippling of the water. With this in mind they watched again, as a fresh breeze stirred the scene. But there was no mistake; the bittern *was* swaying as if being ruffled by the wind. The two observers, now convinced that their eyes were not deceiving them, moved in for a closer look:

> We were within a dozen yards of it now and could see distinctly every mark of its rich, brown, black and buff plumage and yet if our eyes were turned away for an instant it was with difficulty that we could pick up the image again, so perfectly did it blend with the surrounding flags and so accurate was the imitation of their waving motion.

The bittern's remarkable camouflage is possible only because it spends most of its life among the reeds, but when an animal moves between different habitats, it cannot be optimally camouflaged for all of them. In these situations it may be optimally adapted for one background, often where it breeds or where it faces the highest risk of predation. Plovers, for example, are conspicuously marked with disruptive dark bands against their light plumage, but when they are sitting on their open nests, surrounded by pebbles and rocks, they are quite difficult to see. Male and female plovers look very similar, and both sexes usually take turns at sitting on the nest. But for many other birds it is usually the females that incubate the eggs, and their plumage is generally somber compared with the brightly colored plumage of the males. The female mallard (*Anas platyrhynchos*), for example, wears conservative browns and fawns to blend in with the background when sitting on the nest, whereas her flamboyant mate sports an iridescent green head and neck.

Most nocturnal birds, like owls, spend their daylight hours roosting in trees and are therefore well camouflaged to their static background.

Some animals camouflage themselves with foliage and other things, like soldiers in the field. Among the exponents of this habit are several species of crabs, collectively referred to as decorator crabs. These crabs have hooked bristles on their carapace and spend much time collecting pieces of algae, sponge, and scraps of detritus, which they attach to their backs. Some sea urchins similarly adorn themselves with pieces of algae, which are held in place by their spines. The woolly alder aphid (*Prociphilus tesselatus*) covers itself in a fleece of fluffy white wax, secreted from the surface of its body. The brilliant white wax shows up conspicuously against the dark foliage of the alder upon which it usually lives, so this obviously has nothing to do with visual camouflage. The likely reason that they so adorn themselves will not become apparent until we learn something more about aphids.

There are many species of aphids, most of which do not cover their bodies in wax or in anything else. They feed by tapping into leaves and sucking up the plant juices. To obtain sufficient nitrogen from this sugar-rich food source, they have to imbibe large volumes, and they void the excess—a sugary solution called honeydew—as part of their excreta. The honeydew does not always go to waste, though, because aphids are often associated with ants that "milk" them for their honeydew. The ants, in return, provide a protection service, aggressively attacking any interlopers that threaten their milk cows. The story becomes even more intriguing in the case of the woolly alder aphid. This is because of its relationship with the larvae of the green lacewing fly (*Chrysopa slossonae*). These larvae are about the same size and shape as the aphids, which they just love to eat! Given the chance, a lacewing larva will seize a soft plump aphid, pierce it with its sickle-shaped mandibles, and suck it dry within a few minutes, leaving behind a dried husk. But how do they deal with the aphids' private army of ants? The larvae simply disguise themselves as aphids by stripping the fluffy wax off their intended prey and piling it upon themselves. To this end, their bodies are covered in bristles, with hooked ends, to retain the covering. The larvae make such a good job of their camouflage that they fool the ants into thinking they are woolly aphids. Sometimes a "wolf-in-sheep's-clothing" is challenged by an ant, but it usually passes muster.

The investigators who discovered the deception followed the fortunes of twenty-three disguised lacewing larvae that they individually released into different aphid colonies. Eight of the larvae were bitten by an ant after being challenged. But the bites were not serious, and in every case the ant got only a mouthful of wax for its trouble and backed off to clean itself without resuming the investigation. The remaining fifteen larvae were merely inspected, then left to go about their business. The disguise fooled not only the ants but also the investigators, who could only distinguish between aphid and imposter on close inspection, and then only after much experience. In one of their experiments they removed the wax from twenty-seven lacewing larvae by wiping them with a brush. The denuded larvae were then released into the aphid colonies, one at a time, but they did not last very long before being bitten by the guardian ants and dragged away. Sixteen of them were thrown over the edge to the ground below, taking their guards with them in two instances. Seven more were bodily carried off the alder, two of them sustaining injuries during the struggle from which they died. The remaining four larvae managed to escape to an unguarded region of the colony where they took prisoners, stripping them of their wax to restore their disguises.

Aphids offer no resistance when they are being stripped, and the lacewing larvae pass from one aphid to another until they have collected all the wax they need, completing their disguise in less than twenty minutes. In addition to gaining access to a passive herd of prey, the camouflaged lacewing larvae may enjoy the protection of the ants. They probably also escape predation by birds, as well as by other insects, because birds seem to avoid the woolly aphids, which may be the reason that the aphids cover themselves in wax.

A second example of insect invasion—the intrusion of the death's head hawkmoth into honeybee colonies—does not involve the use of an obvious disguise. The death's head hawkmoth (*Acherontia atropos*), featured in the movie *The Silence of the Lambs*, is named for the distinctive marking on its back which resembles a human skull. It is a large moth, with a wingspan of almost 5 inches (12.5 cm), and is distinctively marked in yellow, brown, and black. There is no possibility that it could be confused with the much smaller, and more conservatively dressed, honeybee (*Apis mellifera*). Nevertheless, the moth manages to pass the close scrutiny of the guard bees that protect the entrance to their colony. This is all the more remarkable since

It is not clear how the death's head hawkmoth (left) *is able to elude the much smaller honeybee and gain entrance to the colony.*

the guards will not even admit honeybees from other colonies. The intruding moths are rarely attacked by the bees and do not seem to be much affected by their stings anyway, probably because of their thick cuticle and resistance to bee venom. Once inside the colony, the moths are usually ignored by the bees and move about freely, helping themselves to the stores of nectar and honey. There has been much speculation on how the moths manage to elude the bees. It cannot be through intimidation, because honeybees readily attack the large and dangerous European hornet. One of the remarkable features of the death's head hawkmoth is its ability to make sounds, and it does this using a structure like a larynx, which appears to be unique among insects. The queen honeybee can also make sounds, and these are transmitted through the honeycomb. The sounds of the queen have a profound effect upon the worker bees, causing them to stop dead in their tracks. Since the moth produces similar sounds, it has been suggested that this is how it wins the cooperation of its hosts. However, a recent investigation into the problem failed to reveal any freezing response to the moths by the bees.

Honeybees recognize one another by scent (insect scents are called *pheromones*), so it is possible that the moths gain the bees' acceptance using smell. To test this idea, some investigators extracted the body odor of some honeybees and compared it with a similar extract from the death's head hawkmoth. They found four dominant compounds in the moth's scent that appeared identical to those of the bees. These four compounds were also present in moths that had never visited a bee colony, showing that the moths must have manufactured the scents for themselves, rather than merely picking them up from the bees. The death's head hawkmoth therefore deceives the bees by disguising itself with a scent like their own.

Another example of animals using chemical disguises appears to be provided by a small limpet called *Notoacmea paleacea.* This mollusc, a native of the Pacific coast of North America, is found low on the shore, where it grazes on surfgrass (*Phyllospadix*). This particular marine plant is too flimsy to support the weight of large starfishes, but there is a small starfish, called *Leptasterias hexactis,* which actively searches among its blades for suitable prey. This agile starfish attacks several other species of molluscs but tends to leave *Notoacmea paleacea* alone. When the starfish is on the prowl, the other molluscs become very agitated and try to escape, but *Notoacmea paleacea* shows no reaction. It has been found that the limpet extracts a certain chemical from the surfgrass, which is deposited into its shell, and this may act as a chemical camouflage, making the starfish believe it is just part of the surfgrass.

We have seen that some animals actively camouflage themselves by attaching things to their bodies. Many more animals become passively covered by encrusting organisms, and this may have a similar camouflaging effect, either visual, or chemical, or both. Several species of sponges, for example, grow on various other animals, especially molluscs. In one particular association, between the bread crumb sponge (*Halichondria panicea*) and the variegated scallop (*Chlamys varia*), it was shown that the encrusting sponge greatly reduced the number of attacks by starfishes. This effect seems to result from the sponge impairing the grip of the suckers that the starfish uses to attach to the shell. There are also sea anemones that grow on the shells occupied by hermit crabs. The crabs receive the benefit of being camouflaged by the anemones, while the anemones probably receive small floating scraps of food from the crabs' meals and are therefore in a symbiotic relationship.

We may marvel at a bittern's ability to join in shimmering harmony with the reeds, or the way the death's head hawkmoth can beguile honeybees, but the consummate achievement in animal camouflage must be those species whose bodies can transform into other things. Stick insects are among the most familiar examples, but there is a dazzling cast of other actors, all equally convincing in the parts they play. A common role is to imitate a plant. There are moths (*Draconia rusina*) whose wings are tattered and holed like a partly skeletonized leaf, bugs (*Cycloptera excellens*) whose leaflike wings are blemished as if by fungi, and mantids (*Choeradodis* and *Stagmatoptera*) that look like a clump of leaves. In Brazil there is an Amazonian fish (*Monocirrhus polyacanthus*) that looks like a dead leaf, complete with a barb on its chin to simulate a stalk. It plays the part, too, floating motionless in the water in various attitudes or lying on the bottom. The sea dragon (*Phycodurus eques*), a fish from Australian waters, looks every bit like a floating mass of seaweed, while the filefish (*Aluterus scriptus*) stands on its nose on the bottom of rivers in its imitation of eelgrass. These animals have evolved an external body form and matching behavior to make themselves look less like animals to escape the attention of predators.

The old saying that you are what you eat has special meaning for caterpillars of a moth species (*Nemoria arizonaria*) living in northern Mexico and the southwest United States. Adult moths lay their eggs on several species of oak tree. There are two separate egg-laying periods each year: one in the spring, the other in the summer. Caterpillars hatching from eggs laid in the spring feed on the catkins of the oak trees—those knobbly yellow flowers produced in the spring. Remarkably, the developing caterpillars undergo a dramatic transformation, to look exactly like their food. Their skin turns the same shade of yellow as the catkins, and they sprout bumpy outgrowths, complete with two rows of reddish brown dots along their back to resemble the stamens of the flower.

Catkins are only around for a short period and have long since disappeared by the time the summer brood of eggs are laid. These hatchling caterpillars feed on the oak leaves and start off by looking identical to those of the spring brood. However, as they develop they undergo a transformation to resemble the young twig of the oak. Experiments were performed where eggs from the same female were reared separately, on the two differ-

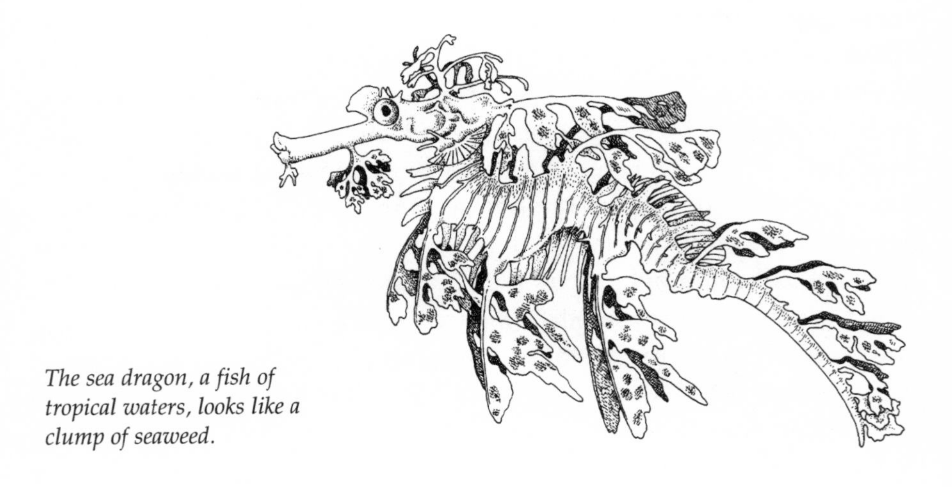

The sea dragon, a fish of tropical waters, looks like a clump of seaweed.

ent food types. Eggs raised on catkins always developed into catkin mimics, while those raised on leaves always became twig mimics. This finding showed that the changes in appearance of the caterpillars were entirely due to substances in their diet. The phenomenon where environmental factors cause the development of more than one body form is called *developmental polymorphism* or *polyphenism*. The developing individual possesses all the necessary genes to develop into one form or the other, but the choice is determined by environmental differences.

An entirely different class of animal impersonators are those that resemble harmful or unpalatable species. This phenomenon, called *Batesian mimicry,* was first described by the English naturalist H. W. Bates. Bates, like several other great Victorian naturalists, spent many years in the tropics collecting specimens, including a large variety of butterflies. And it was while studying these butterflies that he made a startling discovery. Some of the species were almost identical in color pattern with certain other species, even though they were not even members of the same family. What could possibly account for such a remarkable occurrence? Returning to England from his South American odyssey in 1859, he began working on the butterfly problem. He had read Darwin's *Origin of Species,* published in the same year as his return, and sought an explanation in terms of natural selection. Significantly, the species that looked like one another differed in their palatability to birds (their primary predators), one being distasteful, the other palatable. Here was the key to the puzzle. Over the course of thousands of years, the palatable species had evolved, by natural selection, to resemble the unpalatable ones more closely, thereby enjoying reduced attention from the predatory birds.

Although unpalatable species, called models, and the species that mimic them may look very similar, close inspection usually reveals some differences. These differences would not be missed by an entomologist, but it is not important for the match to be perfect. All a mimic has to do is fool a predator long enough for it to make up its mind not to attack. The differences between model and mimic become greater the more distantly the species are related. The unpalatable butterfly *Papilio bootes,* for example, is mimicked by the moth *Epicopeia polydora,* and although they look superficially similar, there are more anatomical differences between them than between an unpalatable butterfly and its butterfly mimic. Incidentally, since moths are mostly nocturnal whereas butterflies are normally active during the day (diurnal), moths that mimic butterflies have to change their lifestyles accordingly.

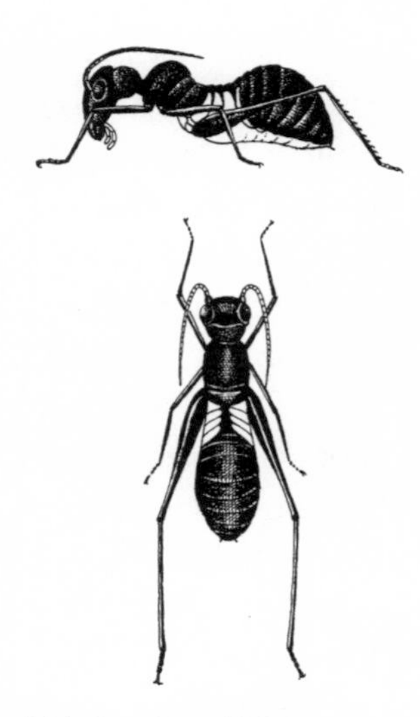

This harmless African grasshopper disguises itself as a stinging ant.

Dissimilarity of body form does not prevent some animals from mimicking others, as illustrated by the harmless African grasshopper *Myrmecophana fallax,* which mimics a stinging ant. Ants have narrow wasp-waists, whereas grasshoppers have much fuller figures. But the grasshopper overcomes the problem by masking out most of its waist with white pigment, making it look as narrow-waisted as an ant.

Most examples of mimicry are found in the insect world, and the idea of finding unpalatable models among "higher" animals, like birds and mammals, appears most unlikely. The 1990 discovery that some birds in New Guinea were highly toxic therefore came as a complete surprise. What was even more astonishing was that their toxin, which was concentrated in the feathers and skin, belonged to a group of compounds (called homobatrachotoxins) hitherto found only in the skin of poison-arrow frogs! Like the poison from the frogs, the bird toxin is highly virulent and causes paralysis. There are three species of toxic birds, all belonging to the same genus (*Pitohui*), and all are brightly colored. What is more, in some regions there is an unrelated bird species whose juveniles resemble the most toxic of the three species, suggesting the possibility of mimicry.

5

Air Strike

Late spring in central Alberta. Meltwater from the winter snows forms a myriad of small ponds beside a permanent lake. They shimmer in the early morning sun. The wetlands are a haven for the waterfowl and shorebirds that flock here in the hundreds. A small formation of teal, perhaps a dozen strong, fly beside the lake, level with the tops of the willows. These small ducks flap their short wings at such a frantic pace that their passage seems labored, but they are strong fliers.

The cloudless heavens are blemished by a small dot, high in the sky, but none of the ducks have noticed it. Half a minute later, the dot resolves itself into a shape that the ducks recognize instinctively. Their tight formation disintegrates in panic. Ducks fall to the ground like stones, as if they had been raked by buckshot. But one teal continues flying on the same course.

The cause of the alarm is a peregrine falcon (*Falco peregrinus*). This crow-sized bird is the fastest predator in the air. The falcon first spotted the teal (*Anas crecca*) while it was soaring at over 3,000 feet (1,000 m) and they were still over 1 mile (2 km) away. Judging the horizontal and vertical distances to its quarry with great precision, the falcon launched itself into a steep dive, called a *stoop*, tucking its wings against its body to reduce drag, thereby increasing its speed. Stoops can be shallow, or as steep as a vertical drop, and may be accompanied by brief periods of rapid wing beating. Estimates of the speeds reached during stoops exceed 200 mph (325 km/h),

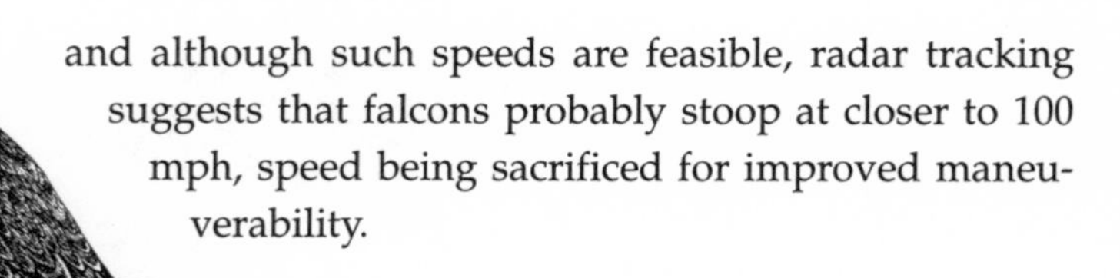

The peregrine falcon

and although such speeds are feasible, radar tracking suggests that falcons probably stoop at closer to 100 mph, speed being sacrificed for improved maneuverability.

The peregrine pulls out of its dive some 30 feet (9 m) above the ground and streaks after the lone teal, wings beating furiously. The duck has a lead of many yards and is flying strongly, but it is no match for the falcon. The gap narrows. Feeling the imminence of the predator's attack, the duck veers sharply to the left in a shallow dive. The peregrine responds by climbing rapidly to gain height, then stoops at its quarry. The falcon's talons are razor sharp and lock securely into the stunned duck's back. Peregrine and teal descend earthward. They land clumsily, and in the brief struggle that ensues the falcon, wings outstretched, straddles the duck. The falcon rips through feathers and flesh with its murderous bill, breaking the duck's neck.

The total distance covered by the peregrine, from the time it first saw the ducks until it made the kill, was about 1 mile, but pursuits may go on for 3–4 miles (5–6 km). Stooping at flying targets is only one of its many attack strategies—the peregrine falcon is a most versatile predator. Birds are its main prey, ranging in size from sparrows and small finches to ducks and gamebirds. Other animals are also taken, when the opportunity arises, mostly rodents like mice and voles.

Observations of peregrine falcons in Alberta showed that their hunting success with small birds, like sparrows and buntings, was lower than with larger birds, like ducks and plovers. This is because the small birds are more maneuverable in the air than the falcon and can effectively dodge away, often to reach the safety of the bush. Falcons therefore frequently stoop down to the ground, often upon flocks of small birds engaged in feeding, hoping to catch one of them unawares before it can fly away. The element of surprise is of cardinal importance, as it is for other predators, especially when hunting fast-moving prey like birds. An alternative to stooping at ground targets is to fly slowly across country at a height of about 30 feet (9 m), eyes cast down in search of prey. Once the prey is spotted, the falcon dashes down to intercept, catching the bird either on the ground or as it tries to gain altitude. Another tactic is to make low-level flights across the land at a height of only about 3 feet (1 m), often at great speeds. The objective here is to flush birds out of hiding immediately in front of the falcon, seizing them as they take off. Peregrines often fly at heights of 250–500 feet (75–150 m) in search of ground targets, descending to lower levels long before reaching their intended target. These flights begin at a leisurely flapping rate but suddenly increase when the attack sequence begins. Alternatively, the peregrine may survey the land from an elevated perch, launching an attack when the opportunity arises, but an active search pattern is its preferred method.

The prey are always seized or knocked out of the air by the falcon's large feet, but the talons, although long and needle sharp, do not necessarily impale them. This explains how small birds sometimes escape, seemingly unharmed, when a falcon accidentally looses its grip on them. The impact of the seizure may be sufficient to stun or even kill the prey, especially if it is a small bird, but killing is usually accomplished with the bill. Falcons have a notch in the upper mandible, which may enhance its cutting ability. While small birds may be so dispatched on the wing, most killings with the bill, and certainly all those of larger birds, are accomplished on the ground. Peregrines spend between about one and two hours hunting each day, depending on food availability. If the food supply is plentiful, they may hunt for less than an hour, spending considerably longer when food is scarce or when they are rearing young. Outside of the breeding season, their hunting often seems quite casual, and they will attack some prey halfheartedly, as if only practicing for the real thing. They may even launch attacks on birds larger than themselves, such as geese, which they would not normally prey upon. But when they get "warmed up," they take their hunting deadly seriously. Their overall hunting success in the Alberta study was about 8 percent, but if their practice runs were discounted (some of which did end in kills), this rate would be much higher. Hunting effort and success change drastically during the breeding season, when they have hungry mouths to feed, with reported success rates in the vicinity of 40 percent. This figure is higher than that for lions but is not in the same league as that for African hunting dogs.

The notch in the falcon's upper bill may enhance its cutting ability.

In most bird species where there is a size difference between males and females (sexual dimorphism), the male is usually the larger and more dominant. The reverse is generally true for predatory birds and is particularly marked in species that hunt the fastest and most agile prey, like other birds and bats. There is no size differential in species that feed exclusively on carrion, like many vultures, or if there is, it is the male that is slightly larger. The discrepancy in size can be quite remarkable. In small hawks, for example, like the European sparrow hawk (*Accipiter nisus*) and the North American sharp-shinned hawk (*Accipiter striatus*), the female is almost twice as heavy as the male. The size discrepancy is somewhat less in the peregrine falcon, females weighing about 65 percent more than males. As a result of the size differential, females tend to hunt larger prey species than their mates. However, during the breeding season the male does all the hunting, supplying food for his mate as well as for the hatchlings, while she stays on the nest.

Peregrine falcons are found throughout most of the world, from the Arctic Circle and the deserts of the Middle East to the tropics of Africa. A number of subspecies are recognized, each adapted to its own particular region. There are many more falcons besides the peregrine, and they, too, typically deploy a variety of attack strategies, like multirole combat aircraft.

Paximada means "hard, dry bread" in Greek. It is an apt name for a pinnacle of rock that thrusts out of the sea, just off the coast of Crete. The island is devoid of freshwater, and the only vegetation is a low chaparral of stunted brush and scrubby thorns. Rising 400 feet (120 m) above the blue Mediterranean, the island is barely 1/2 mile (1 km) long and has little to offer. Nevertheless, it is home to a breeding colony of about 150 pairs of Eleonora's falcons (*Falco eleonorae*). They are smaller than peregrines, weighing about half as much, and with less size difference between the sexes. Being that much smaller, they are more agile, and no bird is too small to be caught. But they are limited at the other end of the size range, seldom attacking birds larger than a dove. They are found throughout the Mediterranean region and extend as far west as the Canary Islands, off the north coast of Africa. The reason the island can support such a large colony of falcons is that it lies on the migratory route for birds moving between their breeding grounds in Europe and their overwintering areas in Africa.

The falcons breed during late July and August, rather than the more usual springtime. This is to coincide with the autumnal migration of birds from Europe to Africa. During the lean times, before the migration begins, the falcons have to feed on flying insects, but once the first waves of migrants begin arriving in the area, there is food in abundance. Both male and female falcons hunt, but when they start breeding, the female falcon spends almost twenty-four hours of her day on the nest, leaving all the hunting to her mate.

Depending on the point of departure, and on prevailing wind conditions, a migrant bird's sea crossing between the coasts of southern Europe and North Africa can vary between a few hours, or less, to about a day. The migrants typically leave at sunset, to complete the crossing at night, thereby avoiding interception by the squadrons of falcons stationed on islands across the Mediterranean. But if the sun rises before a bird has completed its crossing, it has the option of landing on one of the islands and lying low until nightfall. Since the island may be inhabited by falcons, this is not an option that is readily taken. Due to circumstances of geography and weather, the migrants typically enter Paximada airspace during the first few hours of daylight, a fact well known to the falcons.

The male falcon roosts alone, a few yards away from its mate on the nest. They dot the rocks beneath the stars, like fighter pilots awaiting the scramble. Every so often the stillness of the night is broken by the flight call of an unseen male. The calls become more frequent after about 3:30 A.M., and by 5:30, about one hour before dawn, most of the males have taken off. A steady wind usually blows across the sea, and by flying into the wind and adjusting their flapping rate, the males can maintain their station above the island. They spread out on a broad front, keeping about 100–200 yards (or m) apart, at altitudes ranging from just above sea level to about 3,280 feet (1,000 m). The broad net strung across the sky is a formidable barrier to the migrants that have to overfly the island.

The attacks begin at first light, as soon as the migrants become visible to the falcons. Each male marks its own quarry, swooping to intercept when it comes within range. The chances of an attack succeeding are about 11 percent, and if the falcon fails the first time, it often attacks again, sometimes repeatedly. The prey may still escape, but only to repeat the ordeal with another falcon. The migrants often try to get past the island by flying high. Most of them are much smaller than the falcons, and an observer on the ground captures only the occasional flash of white as their lighter underbellies catch the sun. The falcons dive and swoop on them relentlessly, their punishing attacks looking for all the world like a scene from the Battle of Britain. When a falcon succeeds in snatching a bird in its talons, it swoops off to its nest with its prize, stopping only long enough to hand it over to its mate before rejoining the battle. The male will eat later in the day, when the wave of migration has ebbed away. One particularly successful male catches five birds in less than one hour and goes back to catch seven more.

One of the reasons the migrants suffer such heavy losses is that they usually have to run the gauntlet of several falcons in succession. And since many of the attacks occur over water, the migrants cannot seek refuge in the bush. Some migrants do manage to land on the island, though, and provided they find cover quickly and stay hidden for the rest of the day, they have a chance of escaping after dark.

Downtown Toronto. The tree-lined lane that cuts across the campus wears the fresh green mantle of spring. University students amble along the path on their way to classes. Amorous pigeons coo and strut. Squirrels chase each other across the grass. All is right with the world.

Without warning, a dark shape swoops down from one of the trees, falls upon a pigeon in a cloud of feathers, and carries it back to the tree. The assailant then proceeds to pluck and rip the pigeon apart, and devour it, in full view of the startled onlookers. They look from one to the other, unsure of what they have just witnessed. What sort of bird is it? Where did it come from?

The bird, a red-tailed hawk (*Buteo jamaicensis*), is widely distributed throughout North America and is no stranger to the open areas around our cities and towns. However, this particular incident is a little unusual in that these hawks usually prey upon mammals and reptiles rather than birds. Somewhat larger and heavier than a peregrine falcon, the red-tailed hawk is a stockier bird, with broader and more rounded wings and powerful claws. The students witnessed a typical hawk attack. Unnoticed by passersby and pigeons alike, the bird had been perched in the tree for some time prior to the attack, using the high vantage point to survey the scene below for a suitable quarry. It might have chosen to attack one of the squirrels, but the pigeon just happened to be a more promising target at that particular moment. The hawk then leaned forward on its perch, stretched out its great wings, and launched itself in a swooping dive to the ground. By the time the unsuspecting pigeon realized it was under attack, the hawk had struck it a stunning blow, sinking its long talons deep into its body. The pigeon's death was probably instantaneous. Hawks, like eagles and most other predatory birds, have very powerful feet and use their talons for killing, unlike falcons, which use their bills. Red-tailed hawks in the

American West dispatch rattlesnakes with the same ease, puncturing their skulls and lacerating their bodies with their powerful talons. Although hawks do much of their searching for prey while perched in trees and other vantage points, they also hunt on the wing. They soar effortlessly on outstretched wings; theirs is not the energetic flapping flight that typifies the falcon.

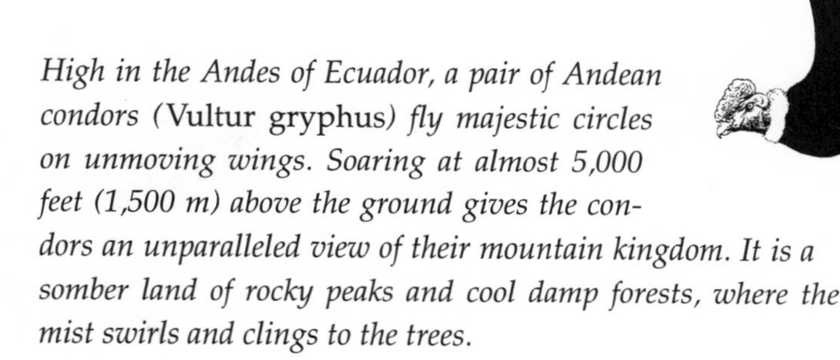

A male Andean condor—the largest flying bird in existence.

*High in the Andes of Ecuador, a pair of Andean condors (*Vultur gryphus*) fly majestic circles on unmoving wings. Soaring at almost 5,000 feet (1,500 m) above the ground gives the condors an unparalleled view of their mountain kingdom. It is a somber land of rocky peaks and cool damp forests, where the mist swirls and clings to the trees.*

Condors are huge birds. The wingspan of an adult male, which is marginally bigger than the female, is about 10.5 feet (3.2 m). This is a little less than that of the wandering albatross (*Diomedea exulans*), whose 11-foot (3.4 m) wingspan is the largest of living birds. But the condor's wings are considerably broader, and it is a much heavier bird. Weighing about 26 pounds (11.7 kg), compared with the albatross's 20 pounds (9.2 kg), the Andean condor is the largest flying bird living today.

"Except when rising from the ground," wrote Darwin, during his 1834 visit to South America, "I do not recollect ever having seen one of these birds flap its wings. . . . I watched several for nearly half an hour, without once taking off my eyes; they moved in large curves, sweeping in circles, descending and ascending without a single flap."

From the ground the condors look like a darker shade of gray against the melancholic sky, but seen up close their plumage is glossy black. They are not handsome birds. Like most other vultures, the head and neck are bare, with loose folds of skin that hang down from the head and throat. The male is further adorned with a great fleshy wattle on top of its head. A ruff of downy white feathers circles the neck like a fur collar, and, except for some white areas on the top of the wings, the rest of the plumage is black.

The North American turkey vulture, an example of a New World vulture.

The African white-backed vulture is an Old World vulture. This group is not closely related to New World vultures, in spite of their similarities.

Turning their bald heads this way and that, they scan the terrain below for signs of their next meal. They have sharp eyes, like other birds of prey, and can pick out small details from a long way off. They have not eaten for two days and are hungry. Several hours of the new day have already been devoted to the search, but so far without any success.

Another hour has passed, but something seems to have caught their attention because they are patrolling back and forth over the same open patch of terrain. They descend for a closer look.

From a height of 2,000 feet (600 m) there is no longer any doubt of the worth of their find—food aplenty for the next several days. Without a single flap of their great wings, they spiral earthward. Pirouetting to the ground with a ballerina's grace, the great birds alight near their prize and prepare to dine.

What had once been a guanaco—a donkey-sized animal like a llama—lies wedged in a rock crevice, legs splayed and broken where it fell. Putrefaction has already begun, and a distended loop of intestine protrudes through a rent in the body wall. There is a strong smell of death and decay, but the two condors, oblivious to the niceties of life, plunge their bald heads deep into the carcass. They pull off great gobs of flesh, bolting it down as if it were ambrosia.

Condors are New World vultures, a group that includes the North American turkey vulture (*Cathartes aura*). In spite of their similarities in appearance, they are not closely related to the Old World vultures, which include the African white-backed vulture (*Gyps africanus*). However, the two groups have similar lifestyles, feeding almost exclusively on carrion. Some vulture species do occasionally kill other animals, but usually only the sick or the young, which offer them little resistance. Almost all vultures have bare heads, often bare necks too, because if they were invested in feathers these would become matted and soiled when they reached inside carcasses. The prospect of eating a decomposing corpse is abhorrent to human sensibilities, but from the vulture's perspective it is the equivalent of our sitting down to a prime rib roast. Vultures have their limits, though, and avoid eating flesh that is too far decayed.

A study of turkey vultures in the tropics showed that they preferred one-day-old carcasses to those that had been dead for four days, though both smelled terrible to the investigator! One of the aims of the study was to find out how vultures locate their food. Although birds generally have little or no sense of smell, olfaction is well developed in some birds, including turkey vultures, which appear to use smell for locating food. To test whether this is true, the investigator set out dead chickens as bait, leaving some out in the open and hiding others from view. The turkey vultures found both kinds with equal ease—provided the carcasses were not too fresh—showing they were using their keen sense of smell for location. The investigator was surprised at how quickly the vultures found the bait, and most of the carcasses were discovered within twelve hours of being left out. The mammalian carrion feeders in the area did not do anywhere near as well, providing no competition for the vultures. This is because flight gave the vultures the advantage of being able to search large areas in a short time, at minimal energy costs.

Condors are graceful fliers, but, because of their large size, they have difficulties becoming airborne, especially when gorged with a large meal. Takeoff requires a long run in which they flap their wings laboriously as they struggle to gain a sufficient airflow over their wings to become airborne. Darwin recounts how the people living in the Andes used to capture condors. They would build a low fence around a carcass and wait for the hapless birds to come and feast. Once the birds landed they could not take off again because the fence prevented them from making their takeoff run. The locals could then club them to death for their feathers.

One of the largest Old World vultures is the lammergeier, or bearded vulture (*Gypaetus barbatus*), with a wingspan of up to 9 feet (2.8 m). Found in mountainous regions throughout most of the Northern Hemisphere, it is primarily a carrion feeder that specializes in bones. One of its unusual habits is splitting long bones open by carrying them aloft and dropping them against suitable rocks. The bird holds the bone in its feet, and if the bone is large it is carried lengthwise, to reduce the drag. Lammergeiers often dive at a rock before releasing the bone, to increase the speed of the impact. If the bone does not break the first time, the process is repeated until it does. Their tongues are stiff and gouge shaped, being specialized for scooping out the marrow from the bones. These vultures also swallow and digest small pieces of bone.

A Scottish moor. A rolling land of bracken and heather as far as the eye can see. The sun, a vermilion orb, sinks behind the distant hills, setting a small stream ablaze in its light. A flight of starlings wheels overhead like a small cloud and flies off toward the east. A few minutes later a lone bird pops up over a low rise in the far distance.

Its effortless soaring flight takes it along the course of the meandering stream. It hugs the ground, flying barely 10 feet (3 m) above the heather, and disappears from view at every dip in the land.

The bird, a hen harrier (*Circus cyaneus*), is about the same size as a peregrine falcon but is more slightly built and weighs much less.[1] Its wings are also much broader, so it has less of a load to bear. The harrier therefore has a lower *wing loading*, which is the weight or mass of the bird divided by the surface area of its wings.[2] Whether a flyer is a bird or an airplane, having a lower wing loading means that it can fly more slowly before the wing stalls. *Stalling* occurs when the air flowing over the wing surfaces—which provides lift—breaks away. The resulting loss of lift causes the aircraft to drop. Jet airliners have much higher wing loadings than light aircraft, and their stalling speeds are consequently higher, about 120 mph (190 km/h) compared with about 45 mph (72 km/h). Being able to fly more slowly makes an aircraft, or an animal, more maneuverable. It also gives a bird of prey, like a harrier, more time to search the vegetation below for potential prey.

The harrier is a courser, flying back and forth across the land close to the ground. Its primary objectives are ground targets, and it uses its long legs to seize small mammals and birds from the vegetation. But it is also adept at snatching birds from the air, especially when they are trying to take off.

The harrier soars across the moor on outstretched wings, intently scanning the ground below for the slightest signs of movement. Cresting a steep rise, it glides down the other side, gaining speed. Razor-sharp eyes catch a small movement in the heather directly ahead, a movement that can be distinguished from the gentle rustle of an early evening breeze. Target or artifact? A split-second decision is made, setting a chain of events into motion. Tail fanned, wings differentially retracted and twisted, the harrier makes a sharp half turn to the left and lowers its long feathered legs. Hanging in the air for several moments, the harrier makes a final adjustment in its position and plummets earthward. Legs leading, talons spread, it disappears from view in the heather.

A full minute passes before the harrier, wings beating powerfully, rises from the ground. Its long legs brush the heather momentarily before being tucked up against its body. Its talons are empty. Whatever had caught its eye either turned out to be of no consequence or had managed to escape. Undeterred, the harrier continues its patrol.

After following the stream for about a mile, the harrier turns around and follows a reciprocal course. The return run is made a short distance to one side of the outward leg, so that the hunter does not go over the same ground twice. But there are occasions when it doubles backs on itself to check a promising location.

Hugging the ground at the approach to a ridge, the harrier sweeps up and over the other side, flushing a pair of ducks from a hollow. The ducks struggle to gain altitude, but the harrier makes no attempt to pursue them. Had the ducks been a pair of skylarks, the harrier would have pressed home an attack, but the ducks are too big and too fast and there is nothing to be gained from tackling unprofitable prey.

Rising out of the hollow, the harrier catches sight of some small movements beside a tussock of grass. Executing a half turn, it hovers in the air for a second or two. And just as it begins its descent, a meadow pipit, the focus of the predator's attention, takes to the air. With a powerful flap of its wings, the harrier throws itself at the small bird. The meadow pipit, being so small, is very agile and almost dodges clear, but the harrier shoots out a leg, catching the wayward bird as deftly as a fielder catching a fly ball. The harrier takes its prize to the ground and rips through its neck with its sharp bill. The harrier flies off, carrying the dead bird in one foot.

. . .

All of the birds treated so far belong to a single order of birds, the Falconiformes, or raptors. Raptors are diurnal and, being animals of the light, locate their prey by sight. Birds in general have good eyesight, vision being their predominant sense, but sight is especially keen among the raptors. We humans have good eyesight too, but it has been estimated that the sharpness, or *visual acuity,* of some raptors may be between two and eight times greater than our own. Visual acuity may be assessed by finding the smallest object that can be seen at a given distance. One way this can be done is to present an observer with a series of gratings, of different mesh size, to see which is the finest one the observer can see. This principle was used in a series of tests on an American kestrel (*Falco sparverius*). The small raptor was given a series of choices between two similar images, one being a fairly course grating and the other a blank. The kestrel was rewarded with food each time it made the correct choice and selected the grating. Once it had learned to perform the test, the experimenters began substituting gratings of finer mesh. In this way they were able to find out the finest mesh size the bird was able to discern. The tests were repeated with humans; the bird's visual acuity was 2.6 times higher. This is the equivalent of the raptor being able to see an object the size of a pinhead (1 mm long) from a distance of about 60 feet (18 m). Although humans did not do as well in the test as the raptor, their performance was not impaired as much as the bird's when the light levels were reduced. Raptors have keen eyesight only during daylight hours. One of the reasons raptors have such good day vision is that they have such a high density of cone cells in the retina (see note 1 for chapter 1 on page 237).

Not only do raptors have sharper eyes than ours, but they can also accommodate much more rapidly, being able to change focus from a distant to a close object almost instantly. This ability is essential for a predator that may close with its prey at speeds in excess of 100 mph (160 km/h). Raptors also have an extensive overlap between the visual fields of their two eyes—they both face forward—giving them a good degree of binocular vision. This feature, essential for depth perception, is a common feature of all predators.

A graphic illustration of raptors' visual superiority over humans was provided by ornithologists Leslie Brown and Dean Amadon. They were standing in a garden in Africa when a buzzard (*Buteo rufofuscus*) swooped down from a distance of 110 yards (100 m) and snatched a small green grasshopper from the lawn. To check their own vision against the raptor's, they put a second grasshopper on the lawn. Even though they knew just where to look for it, and had made it easier for themselves by placing the insect on a contrasting background, one of them was still unable to locate it from a distance of only 35 yards (32 m).

Some raptors are small, and the smallest species, found among the falconets (*Microhierax*), are no bigger than sparrows. But many raptors are large, some having wingspans in excess of 6 feet (1.8 m). As birds get larger, their wing loading increases because their weight increases faster than their wing area.[3] A point is soon reached where flapping flight, which is energetically expensive, is no longer feasible for sustained flying, and soaring flight is used instead. In *soaring,* energy is extracted from the environment—from moving air masses—rather than being supplied by the muscles. The term *gliding* is sometimes used synonymously with *soaring,* but the two are not the same because gliding is a passive mechanism, in which height is continuously lost, whereas in soaring, height is gained or maintained.

One of the most common sources of moving air that soaring birds exploit is thermals, which are rising columns of air. Thermals are often quite narrow, and soarers have to make tight turns to stay within them. Such maneuverability requires slow flying speeds, the same way that turning tight corners in a car requires slow driving speeds. Soaring birds that exploit thermals therefore require lower wing loadings. This is achieved by having broad wings. The relative breadth of a wing is expressed by the *aspect ratio,* which is the length of the wing divided by its breadth. Raptors typically have wings with low aspect ratios. However, the more active fliers, like the falcons and harriers, have higher aspect ratio wings.

The wing tips of most raptors are characterized by their ragged appearance. This is due to the separation of adjacent feathers, like splayed fingers, which results from the feathers being narrower toward their tips. Each separated feather functions as an individual winglet to improve the lifting performance of the wing, both during flight and on takeoff.[4] This high-lift device, called a *slotted wing tip,* has been copied by Boeing and other manufacturers and adapted for use on aircraft as a single winglet. The device is used on 747s and other commercial airliners. Slotted wing tips can be seen on other birds besides raptors, including ravens and crows.

Even the largest raptors have to flap their wings during takeoff, but once airborne they stay aloft by soaring. Aside from exploiting thermals, they use updrafts, as when winds blow against cliffs. I once saw a raven (admittedly not a raptor) high in the Swiss Alps, soaring on an updraft. The bird was only a few feet away from a rock face and was using the

Raptors typically have wings with a low aspect ratio, and ragged-looking wing tips. From top to bottom: turkey vulture, bald eagle, albatross, swift.

The slotted wing tips of raptors, such as this red-tailed hawk (above), *result from the narrowing of its terminal wing feathers* (right). *This high-lift device has been modified as a single winglet, as used on the Boeing 747* (below).

strong vertical current to maintain its station. Condors similarly use the powerful updrafts in their mountainous habitat. The common kestrel (*Falco tinnunculus*), which is about the size of a pigeon, can often be seen hovering over the dikes in Holland on fixed wings. These birds exploit the updrafts of the sea breezes as they blow up against the sea walls. Using the energy from the wind reduces their own energy costs while they search for prey. A more familiar sight, especially in North America, is to see kestrels hovering several feet above the ground in open country while searching for prey. This skittish flight, where the falcon flaps its wings vigorously while remaining stationary, is often referred to as hovering, but it may not be true hovering because that occurs in still air. The kestrel's version appears to take place in moving air: the bird flies into the wind at the same speed as the wind, so maintaining its station. This type of flight has been named *wind hovering* to distinguish it from true hovering flight. One of the singular features of wind hovering is the bird's remarkable stability. Researchers using a high-speed video camera showed that the kestrel's head moves less than $^{1}/_{4}$ inch (6 mm) during flight, regardless of the vigorous beating of its wings. Keeping the eyes still appears to be an important factor in locating, and getting a good fix on, the prey below. Wind hovering is energetically very expensive, and kestrels keep it up for only short bouts of about twenty-five seconds. But the rewards are high, and in one study the kestrels captured an average of almost three prey items per hour—mostly voles—compared with 0.31 when searching while soaring, 0.21 when searching from a perch, and only 0.07 when searching on the ground.

Kestrels, and some other raptors, have the uncanny knack of being able to locate concentrations of voles from the air, even when the rodents themselves are hidden in their underground burrows. Recent work has suggested that kestrels may do this by spotting the streaks of urine that the voles use to mark the trails leading to their burrows. But how could a kestrel possibly see a small streak of dried vole urine from the air? Surprisingly, vole urine is visible in ultraviolet light. Although raptors have not been specifically tested for their sensitivity to ultraviolet light, some other diurnal birds that have been tested are sensitive to it. Other scent trails may also be visible to raptors in ultraviolet light, but that remains to be seen.

Raptors have sharp, curved bills, often markedly hooked, as in eagles. The bill is usually used only for plucking and tearing up carcasses, killing being accomplished by the feet. These are armed with needle-sharp and strongly curved talons (except in vultures), often with one or two cutting edges to facilitate deep penetration. All raptors have powerful feet, and if a large one, like an eagle, landed on your arm, its talons could penetrate to the bone. Falcons differ from other raptors in using their bills for killing as well as for dismembering, and the upper mandible, as noted earlier, has a notch that probably facilitates its slicing action. Raptors also use their bills for plucking feathers or fur from carcasses, prior to feeding.

Not all raptors are predatory. As we have seen, vultures are primarily carrion feeders, and some species live exclusively on the flesh of dead animals. Many of the predatory raptors will also take carrion when the opportunity arises. These include many of the eagles, such as the golden eagle (*Aquila chrysaetos*) and the bald eagle (*Haliaeetus leucocephalus*), birds

The raptor's sharp bill is often markedly hooked, as seen in this Verreaux's eagle.

we usually think of as exclusively predatory. We also harbor many misconceptions about the size of prey taken by the large raptors. Stories have been told of condors carrying off cattle, and eagles flying away with sheep and human babies, but none are founded in fact. We have already seen that condors are almost exclusively carrion feeders, and the largest eagles rarely prey upon animals larger than fawns and adult rabbits. The largest prey taken by raptors weigh about the same as the predator itself—the largest eagle weighs about 13 pounds (6 kg)—which is in marked contrast to some mammalian predators, which routinely attack animals weighing at least twice their own weight. One reason for the discrepancy is that raptors usually have to fly off with their kills, which therefore places an upward limit on their weight. To this end, the wings of raptors are modified for high lifting performance, with highly cambered (arched) profiles, like those of an airliner. High-speed fliers, like swifts and jet fighters, in contrast, have much thinner wings. Another reason for the differences between predatory birds and mammals is that many mammals hunt in packs, enabling them to kill prey much larger than themselves.

Many raptors have cutting edges on their talons to facilitate deep penetration into their prey.

Many raptors, such as falcons, actively pursue their prey, frequently chasing birds in flight. But most raptors are primarily sit-and-wait predators, whether they do their searching from a perch or while soaring on the wing. Those that do their searching from trees can use the foliage for concealment. When a suitable prey comes their way, often another bird, the chase is then just a short jump. Most raptors do not make use of cover, but this has little effect on their hunting success because terrestrial animals—the primary targets of most raptors—rarely look up toward the sky. The prey may get some warning if their attacker's shadow falls across them, but most are probably unaware of the danger until the raptor's talons sink into their body. But death is swift. When a 10-pound eagle strikes a 3-pound rabbit from the air, the impact probably breaks the animal's back.

To see a large raptor like an eagle in action is an impressive sight, but the most striking of them all must be the harpy eagle (*Harpia harpyja*). This

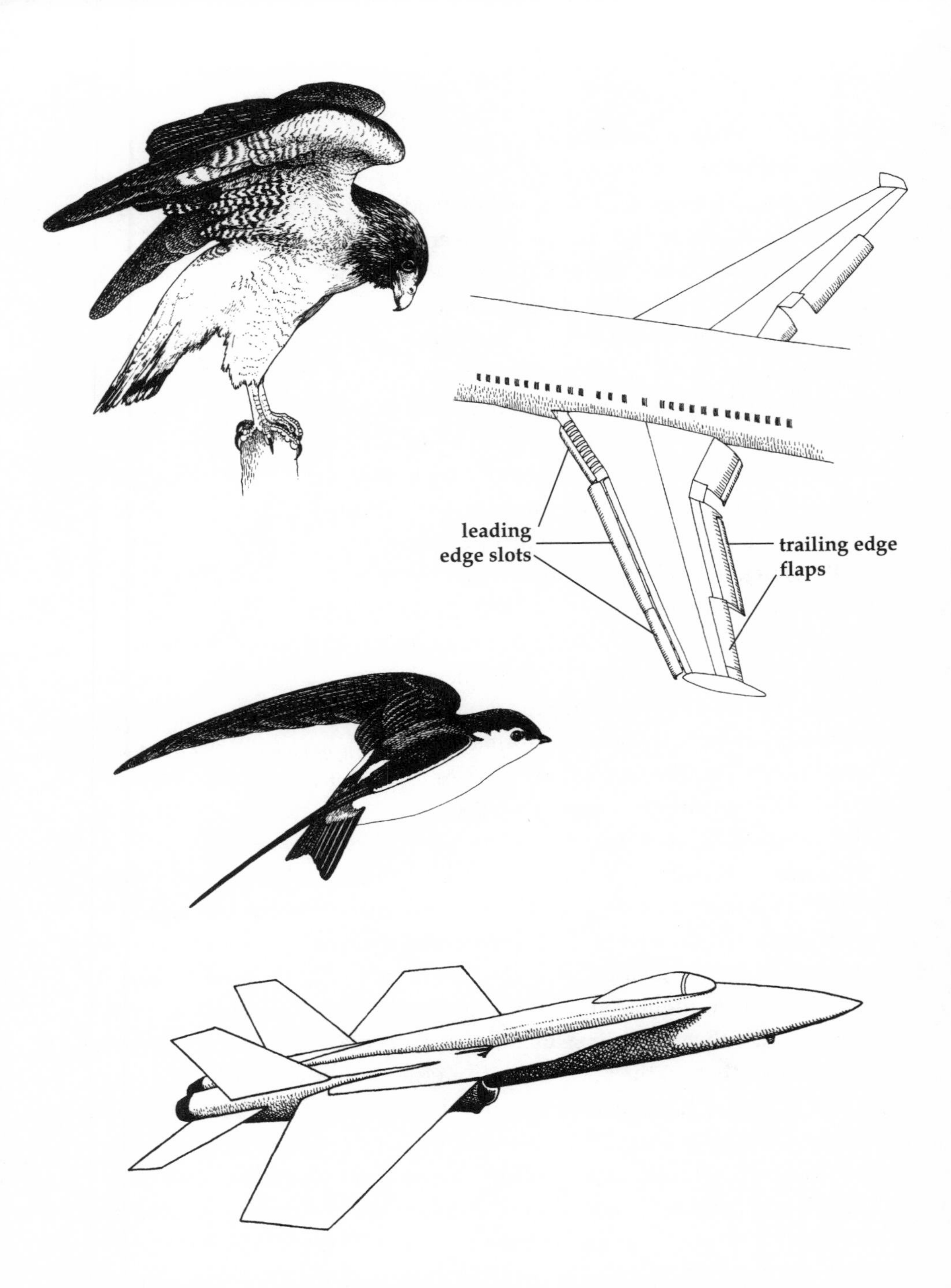

The wing profiles of birds, like those of modern aircraft, reflect their usage. Birds of prey, like this white-tailed falcon (top left) *have thick, highly cambered wings for maximizing lift. Airliners* (top right) *achieve similar ends with their slots and flaps, deployed during take-off and landing. Fast fliers, like the swift* (middle left), *have thin, slightly cambered wings, like high-speed fighter aircraft* (bottom).

powerful bird, a native of the tropical rain forests of South America, is one of the largest predatory raptors. It primarily hunts arboreal mammals, mostly monkeys and sloths, but it also captures birds, including macaws. Remarkably, it hunts below the forest canopy, dodging between the trees in pursuit of its prey at speeds estimated at 40–50 mph (65–80 km/h).

The food requirements of raptors are usually estimated from observations on captive birds, but these can be misleading. Captive peregrine falcons consume about 15 percent of their body weight daily, but since they do not have to hunt for their food, their requirements are less than they would be under natural conditions. Wild peregrines may kill and eat two or three small birds each day, which would amount to more than 15 percent of their body mass. However, they seldom eat the entire bird; they usually eat just the breast meat and perhaps some of the gut. If we took the lower estimate of 15 percent, their annual food consumption would be about fifty-five times greater than their own body weight, compared with twenty times for a lioness. On a weight-for-weight basis, small animals consume far more food than large ones,[5] so the discrepancy between the annual food consumption of a falcon and a lion should be much more, underscoring the shortcomings of using data from captive animals. Although raptors have large appetites in the wild, especially the smaller and more active ones like falcons, their impact on the prey species appears to be small. One of the reasons for this is that raptors hunt over such large areas, because of their powers of flight. The home ranges in which raptors hunt increase with their size. A pair of golden eagles in Scotland, for example, has an average home range of about 13,000–18,000 acres (5,300–7,300 hectares [ha]).

Most raptors are flesh-eaters, but some vultures include some pretty unappetizing food in their diets, like the droppings of other animals. The Egyptian vulture (*Neophron percnopterus*), for example, has been observed in East Africa feeding avidly on piles of lion dung. However, these vultures appear to avoid the dung of wild dogs and hyenas, even though this is just as freely available to them. Could it be that lion dung is more nutritious? This might have something to do with differences in the digestive efficiencies of the different carnivores. A comparison of the feces of felids and canids shows that the former contains less well digested material than the latter. What is more, felids have shorter guts than canids, adding support to the idea that cats do not do such a thorough job of digesting their food as dogs. Why? The zoologist who initiated the investigation suggested it is because of differences in their hunting strategies. Felids are ambush predators and therefore gain from not having to carry the extra weight of a larger gut during their short sprints (see chapter 1). Canids, on the other hand, are pursuit predators, and the extra weight may be of less significance to them during their more leisurely runs. What is of particular interest here is that the same trends appear to hold among predatory birds. Active fliers, like falcons, have shorter guts than less active ones like eagles. The apparent correlation between gut length and hunting style is an interesting one, but it cannot be confirmed (or refuted) until a greater number of species are assessed. Furthermore, it can be argued that it is equally detrimental for a pursuit predator to have to carry a large gut around, given that it probably spends a larger portion of its day being active than does an ambush predator.

A door opens and quickly closes again. It is a cold night and the occupants want to keep every ounce of warmth indoors. The cat, resenting its eviction, slinks across the yard to the barn. The building is old, like the rest of the farm, and the cat has no difficulty squeezing through a gap between its timbers. The barn is unheated, but the warmth from the cows takes the chill off the air. There are also rodents for the cat to hunt in the barn.

The cat's passing does not go unnoticed. The huge pair of eyes that stare out into the night miss nothing. The owl has been sitting on the fence post for almost ten minutes—before that it was perched on top of a derelict tractor. By moving from one perch to another it can search a wider area with the minimum of effort.

The world is an especially dangerous place when you are as small as a mouse. Quite aside from the obvious perils of cats and dogs and other giants that want to eat you is the hazard of being flattened, quite unintentionally, by even larger monsters. The only defense against these horrors is to be unobtrusive and keep hidden from sight. But a mouse has got to eat, and that means coming out of hiding. The safest time to do this is at night, when fewer eyes and fewer big feet are abroad. Not that nighttime is without its dangers. There are all manner of nocturnal predators that eat rodents and other small animals. Nighttime is just the lesser of the evils.

Wings outstretched, body leaning forward, the owl launches itself into the air. It takes one powerful beat with its great wings and soars several yards to another perch. Its passage is completely silent. The new perch on top of a gatepost brings it closer to the barn. It is a clear night, but the moon has not yet risen, and, aside from a narrow chink of light from a shuttered window, the yard is in complete darkness. The owl turns its head through almost a complete circle as it scans the ground below.

A mouse nibbles on a discarded grain of barley against an outside wall of the barn. Finishing its small repast, the mouse sets off on a furtive search for more. Tiny pink feet scurry noiselessly over the frozen cobbles as it picks its way through the darkness. Now the tiny feet are scampering over the desiccated ghost of a maple leaf. The arboreal apparition has little more substance than the air that surrounds it, but it rustles all the same. The mouse stops, sniffs at the leaf for a few seconds, then moves on. The next instant the mouse is dead, and its body is whisked away skyward.

Owls are not closely related to raptors, but they have many features in common, because of their similar predatory ways—owls are *convergent* with raptors. Like most raptors, they have good eyesight, binocular vision, sharp bills and talons, and broad, highly cambered wings. However, they differ from raptors in being primarily nocturnal, though some of them do hunt during daylight. Their eyes are specialized for night vision, but their primary means of locating prey is through their remarkably acute sense of hearing. Correlated with this is their silent mode of flight. The owls are a far less diverse group of birds than the raptors, and an understanding of one species gives a good understanding of them all. One of the most familiar species is the barn owl (*Tyto alba*), which has a wide geographic distribution and a long association with humans. It will therefore serve as a good exemplar for the group.

The barn owl has two hunting strategies: searching from perches, like many hawks, and coursing over the land at a height of about 3–10 feet (1–3 m), like harriers. Its prey consists largely of small mammals, especially voles, shrews, and mice, but it also catches birds, particularly sparrows, and the occasional bat. The prey are seized by the talons, the long legs being especially suited for operating in tall vegetation, but, once caught, the prey are often transferred to the bill for transportation. The prey are usually swallowed whole, but the acidity of an owl's stomach is less than that of raptors, and the bones and teeth are therefore not

Owls have large eyes, and most have a characteristic facial disc—the barn owl's is heart-shaped.

digested. Fur and feathers are not digested either, and the same is also true for raptors. As a consequence, owls cough up pellets containing the undigested parts of their meal, within about eight to twenty-four hours of its consumption—something like a domestic cat regurgitating fur balls. The pellets collect around the owls' perching sites, and since their contents can usually be identified to species, their study gives an accurate picture of what owls have been eating. The pellets of raptors, in contrast, are devoid of bones and teeth.

Most owls have a characteristic facial disc, and that of the barn owl is heart-shaped. The feathers forming the disc are stiffer than the other feathers and serve to reflect high-frequency sounds. Significantly, the two ear openings lie within the facial disc, each surrounded by a depressed area which is continuous with a furrow in the feathers. Sound waves striking the disc are funneled along these paired furrows to the ear openings. Neither the position of the ear openings, nor the shape of the depressed areas surrounding them, are exactly the same on either side. As a consequence, the quality of sound arriving at each ear is slightly different. It appears that this difference increases the owl's ability to locate the source of the sound. Experiments with owls in completely darkened rooms have shown that they can locate mice introduced into the room, provided the floor has been covered with dry material, such as leaves, so that the movements of the mice can be heard. Not only can owls accurately pinpoint the source of sounds, but they can also distinguish between the rustling caused by the wind and that caused by a potential prey. The owl's initial response on hearing a sound is to turn its head so that its facial disc points toward it. When it has obtained an accurate fix on the source of the sound, it swoops down to intercept it. Sounds do not travel very far before being absorbed by the environment or swamped by other sounds. This is why owls do their searching from low elevations, usually no more than about 10 feet (3 m) above the ground. It also explains why they fly in silence, because they do not want to swamp out the noises made by the prey with the sounds of their own wing movements.

I have two wings in front of me. One belongs to a peregrine falcon, the other to a great horned owl (*Bubo virginianus*). When I run my hand over the top of the falcon's wing, it feels quite hard, and the overlapping feathers rustle past one another like stiff paper. But the owl's wing feels downy soft, and the feathers glide past one another smoothly and silently. I cannot see why the owl's feathers are so soft, so I look at them under a microscope. The secret is revealed. The top surface of each feather is covered by a fine

Most owls have soft feathers, and the leading edge of the foremost feather has a prominent fringe. Both features enhance silent flight.

down of minute soft bristles.[6] This soft layer silences the movements of the feathers as they slip past one another when the wings flap. Another striking feature, one easily seen with the unaided eye, is a prominent fringe along the leading edge of the foremost wing feather. It is thought that this fringe smooths the airflow over the leading edge of the wing, preventing turbulence and thereby reducing noise. There is also a fringe on the leading feather of the *alula,* which is the bird's thumb. This small winglet functions as an antistalling device, flicking out during slow flight—usually during landing—to prevent the airflow from separating from the wing. Less obvious than the fringe along the leading edge of the owl's wing are the fringes along the trailing edges of each of the wing feathers. These fringes are also thought to promote *laminar* (smooth) flow, thereby reducing air noise. They probably also help to quiet the movements of the feathers as they move past one another. None of the feathers of the hawk's wing have any of these features. These same features are also missing from fish-eating owls, which locate their prey visually instead of with their ears.

Although hearing is the owl's primary sense, the eyes are also used to locate prey and are modified for operating at low light levels. It has often been claimed that owls' eyes are on the order of ten to one hundred times more sensitive than ours, but this is a wild exaggeration. Humans actually have good night vision compared with most other vertebrates. Investigations on the tawny owl (*Strix aluco*), a strictly nocturnal species, show that its eyes are only about two and one-half times more sensitive than ours. Nevertheless, this modest margin of improved performance can make all the difference between seeing and not seeing an object at night. The owl's increased visual acuity has nothing to do with differences in the sensitivity of the retina. Instead, it is attributed to the construction of the eyeball. As most photographers know, the light-gathering capacity of a lens is expressed by the *f-number,* which is the focal length[7] of the lens divided by its aperture. The lower the f-number, the higher the light-gathering capacity. The f-number of a tawny owl's eye is 1.3, compared with 2.1 in humans, giving the owl an increase of about 2.6 in the amount of light falling on its retina. For reasons I will explain in a moment, the lens in an owl's eye has a high focal length, which tends to increase the f-number. The owl compensates for this by having an extralarge aperture—the pupil is wider.

An owl's night vision may not be very much better than a human's, but a more meaningful comparison is to measure the owl's performance against that of a diurnal bird. Compared with a pigeon, an owl's eye is about one hundred times more sensitive to light, which is a good measure of its superior nocturnal performance.

Not only is an owl's eye more sensitive to light than a human eye, but it also has greater magnification. It is therefore able to see a smaller object at a given distance or the same object from a longer distance. The ability to form larger images on the retina requires a longer focal length lens. This requires a correspondingly deeper eye in which to house it, just as a long focal length camera lens, like a telephoto lens, is much longer than a regular lens. The usual way of achieving such depth is to have a large eye, but this has the disadvantage of adding to the weight of the head. Owls compromise by having a cylindrically shaped eye, giving the necessary depth to house a long focal length lens without incurring too much extra weight. The disadvantage is that the eye has a much narrower field of view, giving the owl tunnel vision. Owls therefore have very little peripheral vision and can only see directly in front of them. And because the eye is so large, it cannot be swiveled around in the eye socket. Owls compensate for these shortcomings by having a particularly flexible neck, allowing them to turn their heads through an arc of about 270°. Raptors also have flexible necks for similar reasons; their eyes are large and have little mobility within the eye sockets.

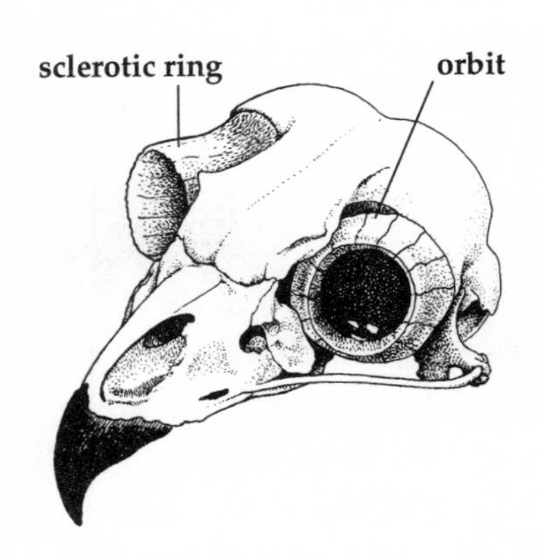

The bony sclerotic ring that maintains the shape of a bird's eye is tubular in owls and is fixed inside the orbit.

. . .

Bats, like owls, are creatures of the night, but most of us, myself included, have had very few encounters with them. Unlike owls, which can sometimes be seen during the day, bats usually leave their roosts only at night, reducing our chances of seeing them. Many bats are small—the size of a chickadee rather than a pigeon—and they are very lightly built. The largest Canadian bat, for example, the hoary bat (*Lasiurus cinereus*), weighs only 1 ounce (30 grams [g]) and is about 5 inches (13 cm) long. Bats are darkly colored, making them difficult to see even when our paths do cross. Our lack of familiarity with bats has fueled our uneasiness and superstitions about them, but they are a fascinating group of animals. One aspect of their biology that has attracted a great deal of attention is their ability to orient and locate prey using echolocation. Their prey are often flying insects, usually moths, which are also creatures of the night. But not all bats are insectivorous, and a wide variety of other animals are eaten, including fish, frogs, small mammals, birds, and other bats. Some bats feed only on plants—fruit, pollen, and nectar. The fruit-eating bats belong to a separate group (the megachiropterans), and, except for the Egyptian fruit bat (*Rousettus aegyptiacus*), they are unable to echolocate. Much has been written about bats, including the very readable *Just Bats* by Brock Fenton, a prolific researcher in the field. I draw freely on his work, and that of his colleague James Fullard, in the pages that follow.

Cetaceans, we have seen, echolocate using a series of clicks, some of whose frequencies are within our own hearing range. Bats, on the other hand, use high-frequency signals for their echolocation calls, ranging from about 8 kHz to a staggering 215 kHz. Theoretically, humans can hear sounds up to 20 kHz, but our sensitivity decreases as frequency increases, and few adults can hear 20-kHz sounds. Most bats use frequencies that lie well beyond our hearing range, so we can only detect their sounds by using electronic equipment. It was not until the 1930s that we had microphones sensitive enough to detect high-frequency sounds, or instruments like the oscilloscope[8] for displaying them. As a consequence, bat sonar remained a mystery to us until that time, though the great Italian scientist Lazzaro Spallanzani (1729–99) suspected from his experiments on bats that they could "see with their ears."

Bats emit their high-frequency sounds as a series of pulses, which may last from less than 1 millisecond (ms; one-thousandth of a second) to about 50 ms. Most bats use short pulses of about 1- to 10-ms duration. The sounds are generated in the larynx, as in other mammals, and emitted through the mouth or nose in the usual way. When the sound impulse hits an object—whether a tree the bat wants to avoid or a moth it wants to catch—it rebounds as an echo that the bat can hear with its ears. By timing how long the sound takes to reach the target and bounce back to its ears, the bat can calculate its range to the object. This calculation requires that the outgoing signal is logged into the brain so that the time delay of the returning echo can be precisely calculated. The bat can also obtain the bearing of the object by detecting the differences in signal strengths reaching its two ears. There obviously has to be a sufficient time interval between outgoing impulses so that a returning echo can be received and analyzed before the next impulse is generated. This interval typically ranges from 5 to 100 ms. By combining short impulses (less than 1 ms) with short time delays, a bat can send out signals at a rate of up to two hundred impulses per second. This is all the more staggering when it is considered that the bat's brain has to make two hundred separate calculations of range and bearing every second. The advantage of using a high pulse rate is that more information can be gathered about the position of the target. This is important if the target is changing positions rapidly, like an insect flying an erratic course to avoid interception.

As we saw in chapter 3, higher frequency sounds have shorter wavelengths and can therefore detect smaller targets than lower frequencies. But as the frequency of the sound increases, so does its absorption by the surrounding air, reducing the strength of the returning echo and thereby reducing the range of the equipment. However, bats can increase their range by increasing the intensity of the sound impulses. At this point I must explain the difference between sound intensity and loudness.

The intensity, or power, of a sound is usually measured in watts. However, the difference in intensities between two sounds is measured in decibels (dB)—that is, the ratio (actually the logarithm of the ratio) of their respective intensities. For example, the difference in intensities between a person talking and an orchestra playing is about 30 dB. In everyday usage we usually use the term *loudness* in place of *intensity,* but there is a subtle difference between the two. Intensity is the power of a sound and can be measured precisely, whereas loudness is a sensation that depends on the observer. As most parents will attest, a teenager's perception of loudness differs from that of an adult's. A teenager may try and convince her parents that the stereo is not loud, but any agreement on what constitutes "loud" is unlikely to be reached. However, it is possible to measure the sound intensity in the room where the stereo is blaring; it might be 90 dB higher than when the stereo is unplugged. Although intensity and loudness are not the same, it is generally true that loudness increases with intensity. So there is no great harm in using the more familiar term *loudness,* provided we remember that it is subjective and imprecise. For example, some bats, as we will shortly see, generate high-intensity sounds. We might refer to them as loud, and they are loud from a bat's perspective, but we cannot hear the sounds (the frequencies are too high) so they are not loud to us. When I use the word *intensity* in the sections that follow, make a parenthetic note that it essentially means "loudness."

The echolocation sounds of bats can be remarkably intense—in excess of 110 dB just a few inches (10 cm) in front of the bat's face. This is more intense than a domestic smoke detector. If bats are making such loud sounds, emitting them at rates of many times a second, how do they stop themselves from going deaf? Bats have two ways of doing this: neurally and mechanically. In the first method, they reduce the strength of the nervous impulses coming from the *cochlea,* the structure in the inner ear that translates sound waves into nerve impulses. In the mechanical method, they dampen the movements of the three small bones in the middle ear that transmit vibrations from the eardrum to the cochlea. In the little brown bat (*Myotis lucifugus*), for example, the three bones are disconnected during the emission of the sound signal. Although some bats use high-intensity sounds (about 110 dB), others use intermediate ranges (75–100 dB), as loud as a noisy stereo. Some species, often referred to as whispering bats, use sounds as quiet as 65 dB, as loud as a raised voice, giving them a short range of less than 2 feet (0.5 m). The maximum range of any bat is probably about 30 feet (9 m).

Some bats emit impulses at a fixed frequency, but most others change the frequency during the transmission, starting off at a higher pitch and dropping down. The big brown bat (*Eptesicus fuscus*), for example, starts off at 80 kHz, dropping down to 30 kHz, while the little brown bat sweeps down from 100 to 40 kHz.

The decrease in range as the frequency of the impulse increases is also accompanied by a narrowing of the beam. High-frequency sounds therefore give a narrow beam, suitable for detecting objects at short range with great precision. Lower frequencies give a broader beam, suitable for detecting objects at a greater distance, and on a broader front, but with less precision. By combining the two frequencies in a frequency modulated (FM) signal, the bat is able to combine long-range detection over a wide area with more precise short-range detection over a narrower area. But bat sonar gets far more sophisticated than this because bats can change their signals as they close with their target. Bats that use FM signals shorten the impulse times as they close with their prey, making the drop in frequency much steeper. This gives them more precise location data. And as they get closer to the target, they emit two sweeping sound signals at the same time, which gives them a precise fix on its position. But there is yet another level of complexity to bat sonar, and this pertains to the Doppler effect.

When an object that is emitting sound moves toward an observer, the sound waves become compressed, causing the wavelength to decrease and the frequency to increase correspondingly. The reverse happens when the source of sound is moving away. This is called the Doppler effect, and we experience it whenever fast-moving noisy objects, like fire trucks and trains with sirens, approach us. When such a vehicle is speeding toward us, the pitch of its siren sounds higher than it is. But when it goes past and starts moving away, we hear a sudden drop in pitch. The people traveling in the vehicle hear no change in pitch.

The sonar system of bats takes account of the Doppler effect when calculating distances to a moving target. Even more impressive is the fact that some bats can exploit the Doppler effect, and in two different ways. First, by changing the frequency of their outgoing impulses, some bats can adjust the altered frequency of the returning echoes so they fall within the most sensitive part of their hearing range. Second, some bats can detect the fluttering movements of a flying insect's wings by the Doppler effect they cause. As the insect's wing tips move toward the bat, they increase the frequency of the bat's sonar impulse, and the reverse occurs when the wing tips move the other way. These alterations in the frequency of the returning echo are interpreted by the bat as representing an insect, and it is even possible that some bats can identify their prey from the characteristics of the Doppler shift.

Although bats typically search for flying insects on the wing using their sonar, some species adopt a sit-and-wait strategy, using their eyes to search the ground for suitable prey. The heart-nosed bat (*Cardioderma cor*), for example, sometimes perches about 3 feet (1 m) off the ground, swooping down on crickets, scorpions, centipedes, and any other tasty invertebrate that catches its eye. When the fringe-lipped bat (*Trachops cirrhosus*) is hunting for frogs, it often switches off its sonar and locates them by listening for the mating calls of the males. The sonar is then turned on again when the bat closes with its target. Similarly, the Indian false vampire bat (*Mega-*

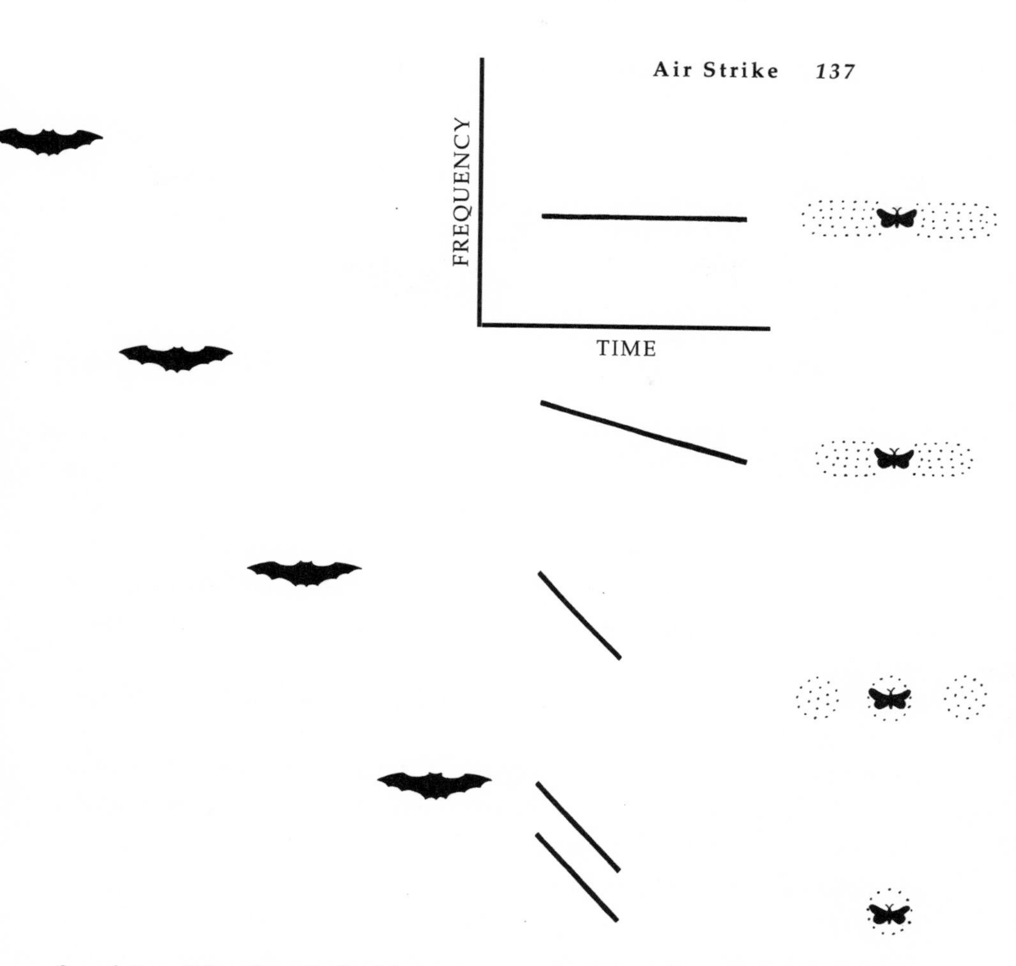

Some bats emit impulses of a fixed frequency, as depicted by the horizontal line (top). *Most other bats change their frequency, starting high and sweeping down during the impulse, as depicted by the sloping line* (second row). *The higher frequency impulses give a narrower beam, as depicted by the dots, giving greater precision in detecting targets. A wider frequency range and shorter impulse time, as depicted by the steeper line, gives a more precise fix on the target* (third row). *From the bat's perspective the location of the target has been narrowed, but there is still some ambiguity in the exact position of the target, as depicted by the three narrow zones of dots. Emitting two sweeping sound signals at the same time gives a bat a precise fix as it closes with its prey* (bottom), *eliminating all uncertainty.*

derma lyra) often turns off its sonar when hunting for mice, using the sound of the prey's footfalls to locate them. The reason bats tend to switch off their sonar when hunting other mammals is that many mammals, rodents included, can hear high-frequency sounds. Another reason for bats to turn off their sonar and use other cues is when they are hunting in a "cluttered" setting that interferes with their sonar signals. In these situations they may use vision or the sounds made by their prey. Bats have exceptionally large

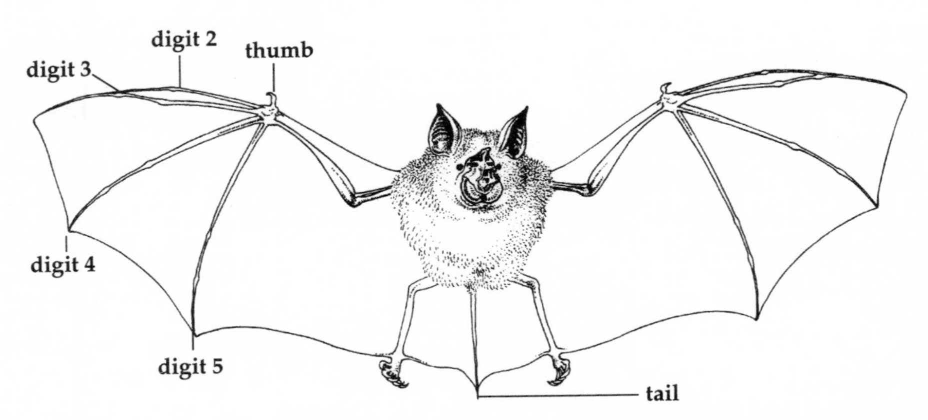

A horseshoe bat, showing how the membranous wings are supported by the fingers and legs.

ears, which amplify the incoming sounds and enhance the bat's directional hearing abilities.

Some bats, including the Mexican bulldog bat (*Noctilio leporinus*), consume quantities of fish. Their sonar system is so sensitive that they can locate a fish swimming just beneath the surface by detecting the ripples caused by its projecting fins. These bats have long toes with sharp claws that they use to gaff the fish. Bats that prey on other bats listen for the characteristic sonar frequencies of their intended targets, just as attack-class submarines detect their targets by identifying the characteristic underwater sounds made by their screws.

The bat's sophisticated target-locating system is supported by its outstanding flying performance. Its wing loadings are much lower than those of comparably sized birds, conferring greater maneuverability. Further enhancing maneuverability is the way the wing membrane is supported by the fingers as well as by the legs. This confers greater versatility to the shape of the wings, which can be constantly adjusted during flight. Many bats also have a separate membrane between their legs that not only contributes to lift and flight control, but also provides a scoop for catching prey, which are then transferred to the mouth. The wings can also be used for catching prey during flight, again transferring them to the mouth. Insects taken from surfaces, as when a bat lands to catch its prey, are usually caught in the mouth. Bats have sharp teeth, and some of them have a phenomenally fast chewing rate—seven times a second for the little brown bat—which enables them to eat large numbers of insects during a feeding foray.

With such formidable aerial predators lined up against them, what chance do flying insects have against bats? A large number of insects, including numerous species of moth, have ears, and these are particularly sensitive to very high frequency sounds like those used by echolocating bats. Significantly, moths collected in a given locality are more sensitive to the frequencies used by the bats in their area than to any others. And it appears that some moths have ears that are more responsive to bat echolocation frequencies than our most sensitive bat-detecting microphones. Some species of moths can detect bat sonar from distances of up to 130 feet (40 m), which is at least twice as far away as bats can detect them. The reason bats are at this disadvantage is that their sound signal has to travel to the target and back again—twice the distance between bat and moth. Bats that use high-intensity impulses obviously give their potential prey an earlier warning of their presence than bats that use lower intensities, which may explain why some bats use quieter impulses. Other bats overcome the problem by using different frequencies from those of the other bats in the area. The trident bat (*Cloeotis percivali*), for example, which feeds extensively on moths, uses an unusually high frequency of 210 kHz, at an intermediate intensity level. Hildebrandt's horseshoe bat (*Rhinolophus hildebrandti*), on the other hand, which feeds on a variety of insects, of which moths make up only a part, uses a more common frequency of 50 kHz, at a higher intensity.

When a moth detects bat sonar its response depends on the closeness of the bat, as judged from the intensity of the bat's echolocation calls. If the bat is a safe distance away, too far to have detected it, the moth simply flies away from the source of the sound. But if the bat is close enough to have locked onto it, the moth starts flying erratically in an attempt to give the bat the slip. One last line of defense is for the moth to stop flying and land—a strategy that we saw used by ducks when under attack by falcons. But some moths appear to have evolved a way of jamming the bat's avionics equipment by emitting high-frequency sounds of their own. One of these is the dogbane tiger moth (*Cycnia tenera*). Its strategy is to remain silent until a bat has obtained a sonar lock on it and is in the final stages of its approach. The tiger moth then emits a series of clicks, similar in intensity and frequency to those that the bat is expecting to receive from its target. This apparently has the effect of immediately disrupting the bat's data-processing system, causing so much confusion that the bat aborts the attack and veers sharply away from its prey. Alternatively, the moth's clicks may simply startle the bat, causing it to abort its attack. It is also possible that some moths use their clicks to warn bats that they are distasteful.

If the attacking bat has had previous experience with unpalatable moths that have emitted a warning signal, it is likely to break off its attack. These three alternatives—sonar jamming, startle, and warning—are not mutually exclusive, and it may be that different species of moths have different strategies.

I stand in awe of the complexity and sophistication in the animal world. The tiger moth's sonar-jamming strategy is all the more impressive when you think that its own data-processing equipment, its brain, is not much bigger than a pinhead. But insects, as we will see in the next chapter, are full of surprises.

6

The Scorpion's Sting

The 1950s inspired some of the worst science-fiction movies ever unleashed on innocent audiences. *The Incredible Shrinking Man* may not have been the worst of the bunch, but it was not one of my favorites. The hero of the movie becomes enveloped in a radioactive cloud while he is out boating. The effect of this exposure is that he starts getting smaller, eventually shrinking to the size of a garden insect. By the use of giant props, the actor is made to look convincingly diminished in size, but there is more to being small than having to use two hands to pick up a pin! The biological implications of being an inch tall are enormous, and I cannot blame the director of the movie for not getting into them, but it would have been fun. If it were possible to scale ourselves down to the height of a matchstick, and have us survive, what would it be like? Aside from such physiological differences as taking hundreds of breaths per minute and having a heart rate of about one thousand beats per minute, what could we expect? One obvious difference is that we would be considerably stronger and more athletic than we had ever been before. If the rest of our world had been scaled down to the same extent, we would be astonished at what we could achieve. Jumping onto the roof of a house would be simple enough, and we could jump down to the ground again without any fear of hurting ourselves. We could carry an object several times our own weight—say a piano—for a city block without a second thought. And if we wanted to, we would probably be able to keep up with a car driving along the highway. The reason for all these remarkable changes has to do with the relationship between areas and volumes as things change size.

As things get smaller, their areas become progressively larger relative to their volumes. This increase in area-to-volume ratio has profound consequences for living things, as well as for inanimate systems, and our everyday lives abound with examples. The peas on a plate cool down far more rapidly than a baked potato. This is because heat is lost through surfaces, and a pea, being so much smaller than a potato, has a much larger area-to-volume ratio. If we want to cool the potato down, we do so by cutting it into smaller pieces, thereby increasing its area-to-volume ratio.

The small size of babies makes them lose heat far more rapidly than adults, which is why they are kept well wrapped. Incidentally, one of the reasons babies have to be fed so often is that, being smaller, they have higher metabolic rates and therefore higher food requirements. This helps explain why they breathe more rapidly, and why their hearts beat so much faster than an adult's.[1]

The correlation between muscle bulk and strength is self-evident from watching weight lifters. People with bulging muscles can lift heavier weights than those of lesser build because the force that a muscle can exert increases with its cross-sectional area. Relative areas, including cross-sectional areas, increase as things get smaller, so strength increases accordingly. On a weight-for-weight basis, small weight lifters can lift heavier weights than large ones. For example, in the men's 242-pound (110 kg) body-weight class, the record stands at 1,047 pounds (475 kg), which is 4.1 times the champion's own body weight. However, in the 114-pound (52 kg) category, the record stands at 5.2 times the holder's body weight.[2] The difference in performance is much more impressive if larger size differences are considered. When a 4-ton Indian elephant picks up a 2,000-pound (910 kg) tree trunk, it seems impressive, but this represents only about 22 percent of its body weight. This is equivalent to my carrying a 36-pound (16 kg) load, which is unimpressive. I am not very strong, but I can carry 100 pounds of plaster when I am in the field, though not for very far. This is equivalent to about 60 percent of my body weight, which is far greater than an elephant can achieve. But this is nothing compared with the performance of small animals like insects. Leaf-cutting ants, for example, can carry in their jaws a piece of fresh leaf that is much bigger than themselves, and when a dung beetle rolls one of those unsavory spheres across the African countryside, it is probably equivalent to our pushing a car.

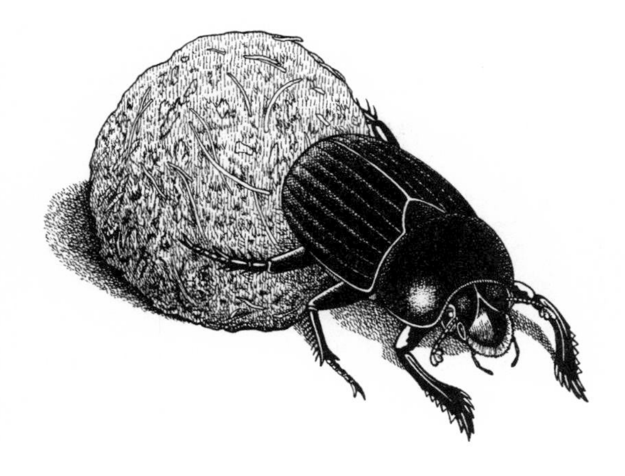

Dung beetles can move mounds many times their own weight.

We step on ants most summer days without even noticing them. And even if we did take time out from our busy schedules to pay them any heed, I doubt they would impress us. But if we were closer to their size, we would take a profound interest in them. And they would take a profound interest in us, but for entirely gastronomic reasons. Insects and their arthropodan (joint-legged) kin live in a world so far removed from our own as to belong to a different universe. If we could enter that universe, we would encounter beings as alien and terrifying as the best science-fiction movies—a universe where the difference in scale makes nonsense of our concept of physical reality.

There are almost nine thousand species of ants. All are colonial, like termites and some bees and wasps. They live together in large numbers, sharing the work of the colony. The workers are sterile females, and each is armed with a sting at the end of the abdomen. A common feature of colonial insects is the caste system, where different individuals are specialized for performing different tasks. This leads to modifications in their bodies, described as *polymorphism* (many forms), and more than two dozen castes of ants have been recognized. The various castes within a colony often look markedly different from one another, both in size and shape. Some of the specializations are quite bizarre, like the replete caste of the honeypot ant (*Myrmecocystus mimicus*). Members of this caste serve as storage containers, and the other workers keep filling them up with the nectar and honeydew they have collected, until their abdomens are grossly distended. The living storage vessels then cling to the roof of the gallery—part of the living space of the colony—their spherical abdomens hanging down like Chinese lanterns.

Ants accomplish their biting and cutting with their mandibles, the

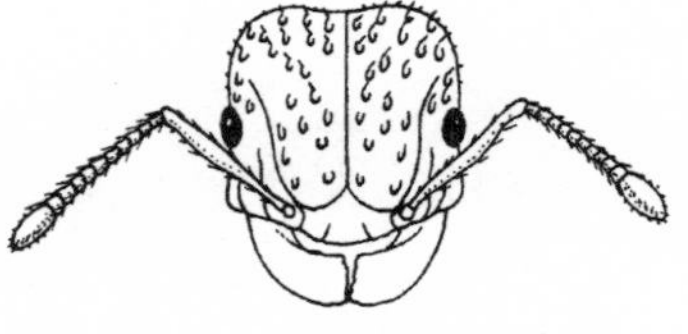

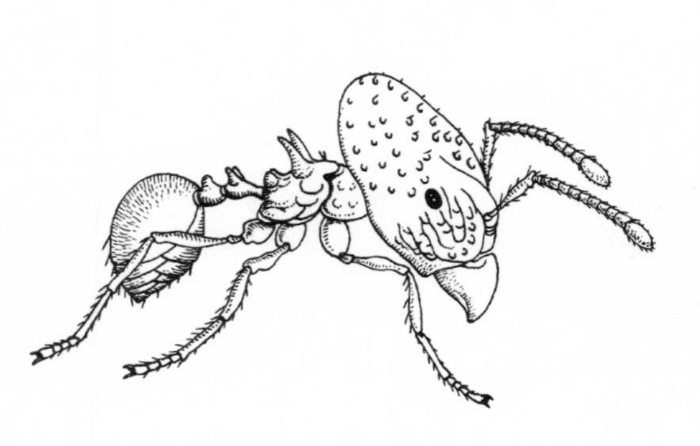

Individual ant castes are specialized for certain roles: the honeypot ant serves as a living storage vessel (bottom)*, while the major worker of* Acanthomyrmex *is a seed miller.*

pincer-like structures that lie immediately in front of the mouth. There are all manner of specializations here, from the short crushing mandibles of *Acanthomyrmex,* used for milling seeds, to the murderously long snapping mandibles of *Orectognathus* and *Strumigenys.* These elongated jaws, which are armed with terminal teeth, are pulled back before use, like a spring trap. A pair of long sensory hairs project forward between them, functioning as range finders. The ant moves slowly toward its prey, and when the sensory hairs make contact with the unfortunate individual, the ant snaps its mandibles closed. The teeth at the ends of the mandibles impale the prey, puncturing its body. Depending on its size, the prey may then be raised off the ground. If necessary, the ant swings her abdomen between her legs and stings the captive. The combination of impalement and injection with venom soon kills the prey.

Ants are mostly diurnal, workers setting off on foraging expeditions first thing in the morning, while the other workers attend to their duties about the colony. Many species feed on plants, but most of them are predatory, preying on other ant species, and upon other small invertebrates. Individuals may go off in search of food on their own or in groups. They may either find their own way or follow the fixed pheromone (scent) trails marked on the ground by previous foragers.[3] Pheromones are used for a variety of communication functions among insects, and many other arthropods, from advertising for mates to marking sources of food. Pheromones are species specific, like the songs of birds, and are therefore reacted to by

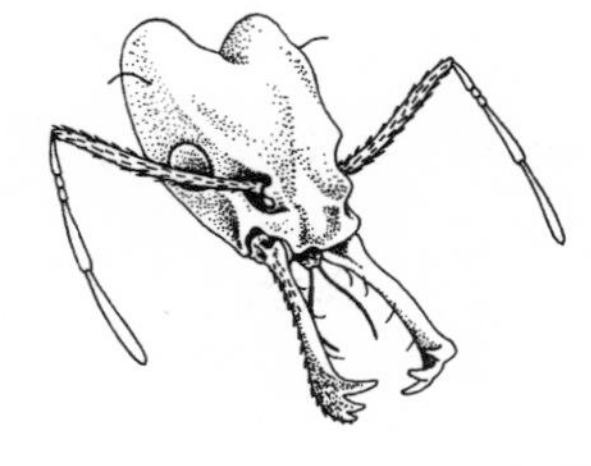

members of the same species. The foraging expeditions are often of short duration, perhaps half an hour or less, though they can be longer, and one individual will usually make several forays during the day.

Removed from the safety of the colony, the lone forager is at considerable risk of attack from a wide range of predators. These include other ants, spiders, predatory flies, lizards, toads, and birds. Some of the lone foragers, like *Cataglyphis bicolor,* can run remarkably fast, apparently up to 1 yard (1 m) per second. This is the equivalent of a lion running at several hundred miles per hour! But in spite of their speed, the foragers have very short lives—less than a week for some of the colonies that have been studied. Although having a short life span may be unfortunate from the individual's point of view, it is of no consequence to the well-being of the colony. During the worker's short life it will have collected food equivalent to fifteen to twenty times its own body weight, which justifies its existence in terms of the economy of the colony. And new workers are bred at a sufficient rate to sustain their heavy losses.

The ant Orectognathus *has murderously long mandibles, shown cocked ready for closing (*bottom left*) and snapped shut (*bottom right*).*

Ants of the genus *Myrmecia,* collectively referred to as bulldog ants, have long mandibles of the spring-trap variety. They ambush their prey, and individuals have been observed in eucalyptus trees, crouched low against the branch for as long as half an hour, with their jaws cocked in readiness. One such individual was seen to catch a bush-fly (*Musca vetustissima*) that had landed about 1 inch (2.5 cm) away from it. These flies are fast, but the ant sprang at it, seizing the fly's head in its mandibles and stinging it to death. The ant then leaped down to the ground, 1 yard (1 m) below, which is like our jumping off the top of a high-rise apartment building.

While the individual exploits of solitary foragers are as varied as they are impressive, few spectacles compare with the mass attacks waged by army ants. There are many species of army ants, a broad term which includes driver ants and legionary ants. These ants vary in the details of how they wage their wars, but they all have two elements in common: the mass

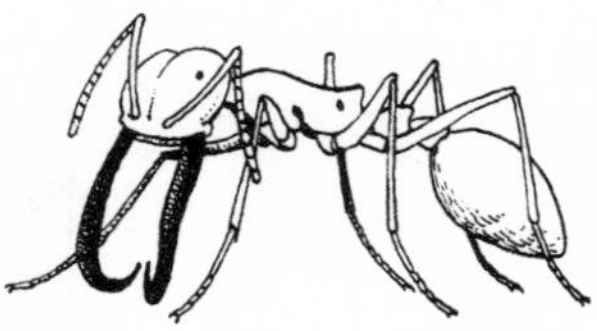

A worker (top) *and soldier caste of the army ant* Eciton.

attack of prey, and the migration of the colony. Some army ants, like *Eciton*, cover large distances after abandoning their old nest site, marching every day and making temporary campsites each night. Their bivouacs are not elaborate affairs. They choose a spot beneath a fallen tree or between the flangelike buttresses that form at the base of certain tropical trees. Their primary concern is to provide protection for the queen and the immature individuals, and they do this by forming a living mass of workers, linked by their legs. The living wall of between 150,000 and 700,000 ants is up to 1 yard (1 m) in diameter.

The bivouac begins to disintegrate at first light. A jostling crowd eventually forms up on a broad front, up to 15 yards (15 m) wide, and begins moving out. There are no leaders, but the smaller and medium-sized workers head up the advance. They run ahead for a few inches, laying down a pheromone trail as they go, then run back to the main body while others take their place. The larger, less agile soldier caste, with their large heads and long sickle-shaped mandibles, fall in at the sides. The column advances at about 20 yards (20 m) an hour, attacking and killing everything in its path. The column of marching ants makes a characteristic noise, partly a rustling of the undergrowth and partly a buzzing and scampering of arthropods as they try to escape. There is also the occasional call of the antbirds, which accompany the column like gulls behind the plough, swooping down to feast on the flushed-out insects. Large and small animals alike are killed, and if a python, heavy with a recent meal, is unfortunate enough to find itself in their path, it will be devoured, alongside tarantulas and other lesser beasts. It is not the sting of the army ants that does the damage—they seldom use it anyway—but the bite from their mandibles.

Army ants of the genus *Dorylus* undertake migrations lasting several days, after which they bivouac beneath the ground. They remain underground for a period ranging from one week to two or three months. A raiding swarm sets out from the secure underground encampment on most days, advancing as a narrow column several million strong. The ants are so numerous that they clamber over one another, spilling out at the sides as the living river flows across the land. They advance at a similar rate as the previous genus and are just as destructive to everything in their path. If a house stands in their way, they will sweep right through it, killing livestock and pets and eating meat out of the kitchen. According to a nineteenth-century explorer, criminals have been executed by tying them up in the path of an advancing column of ants.

During my first visit to Galápagos, I climbed up to the rim of a volcano to look at the giant tortoises in their natural habitat. On the way down, our party came across a small group of tortoises that had stopped to drink at a small stream. I settled myself on a nearby rock to watch them, but within a few moments I was on my feet with my shorts on fire! The cause of my discomfort was fire ants (*Solenopsis*). The sting of these tiny ants is out of all proportion to their diminutive size. The burning sensation, for which they are so aptly named, is attributed to a class of compounds (alkaloids) in their venom called piperidines. The effects of the sting I received lasted for several hours, making me very cautious of where I sit down on the rare occasions that I visit the tropics. Fire ants, which are native to the tropics, have spread to other parts of the world, including North America, probably as unwitting stowaways on imported fruit. They are predatory on other insects and have become something of a pest in the southern United States. They also prey upon vertebrates, especially young individuals, which are vulnerable to their stings and bites. A five-year study of cottontail rabbits in the United States, for example, revealed that over one-quarter of their litters were destroyed by the fire ant species *Solenopsis saevissima*. Pesticides have failed to control them, but they have a natural enemy in their Brazilian homeland in the form of a small fly (*Pseudacteon solenopsidis*). The fly commences its attack on the fire ant by hovering less than 1/2 inch (1 cm) in front of it. Being so small makes the fly very maneuverable, and when the ant tries to run away, the fly stays in front of it by flying backward. Choosing its moment carefully, the fly makes a sudden 180° turn and stabs the ant through a small hole at the back of its head. But instead of injecting venom, the fly inoculates its prey with eggs. When the eggs hatch out, the larvae eat the ant alive. It has been suggested that the fly could be used to control the fire ants in North America. The strategy of laying eggs inside a living host is used by a large number of other insects, including the ichneumon wasps.

There are many more examples of ants and their predators, but I will limit myself to one last case. This concerns a lizard (*Acanthodactylus dumerili*) that lives in the Saharan desert, where it preys on a particular ant, the

Saharan silver ant (*Cataglyphis bombycina*). During the summer, noonday temperatures in the Sahara can exceed 60°C (120°F), and most animals avoid the searing heat of the midday sun by taking shelter in burrows. But the ants, like mad dogs and Englishmen, choose precisely the hottest part of the day to forage for food. They are scavengers, feeding on the bodies of animals that perished in the sun after failing to return to their burrows following their nocturnal feeding forays. And the reason the ants risk a similar fate has to do with the behavior of the lizards.

The lizards, which frequently burrow next to the ants, can tolerate high temperatures. They therefore hunt at the height of the day, seeking out ants because they are the only prey then available. But the ants can tolerate slightly higher temperatures, and time their excursions to the surface to coincide with when the lizards have been forced to return to their burrows. It is impractical for the ants to monitor the lizards' movements, so they choose their moment by reference to air temperatures. Scouts are sent to the surface around midday, and when the air temperature reaches 46°C (116°F), they inform the others. There follows an immediate exodus from the ants' burrow. Because of its small size, the ant's body warms up and cools down very rapidly, so its temperature is very close to the air temperature at that particular point. Ambient temperatures are not uniform, being highest on the surface of the sand and decreasing with height. The ant's lethal temperature is 53.6°C (128°F), giving them a narrow thermal window of only 7°C (12°F). When ants encounter temperatures close to their lethal level, they have to seek lower temperatures to cool off. They do this by climbing up the stems of plants. If they are unable to find such refuge, they die. The ants therefore place themselves at great risk of perishing in the intense heat of the noonday sun. But in so doing they reduce the risk of falling prey to the lizards. Experiments have shown that if ants are released onto the surface prematurely, they are soon snapped up by the lizards.

Termites, often referred to as "white ants," are very distant relatives. They resemble ants in being colonial and in having a caste system. Like ants, they are numerous, and it has been estimated that the two groups account for about three-quarters of the total insect biomass in tropical rain forests. Termites feed largely on wood and are therefore not in direct competition with ants for food, but they do compete for nest sites in fallen trees and in leaf litter. The two groups have coexisted for 100 million years, but it has not been a peaceful relationship because of the intense rivalry between them. Ants, being mostly predatory, are the primary aggressors in the wars waged between the two factions, and the termites have evolved an impressive weaponry for defending themselves. Most ant species, given the chance, will prey upon termites, but some make a specialty of raiding termite nests. One of these is the African species *Megaponera foetens.* Led by a single scout that marks the way with a scent trail, a column of ant raiders two or more workers wide sets out for a termite colony. On arriving, the column breaks up, and the individuals storm the termites' stronghold, squeezing through every hole and crack they can find. Each invader makes its way through the galleries, seizing one or more termites, which are dragged to the outside. The badly wounded prisoners are left to squirm while the ant goes back for more. When each ant has collected about half a dozen termites, the column regroups, the victors pick up as many of the prey as they can carry, and they march back to their own colony.

Soldier castes of various termites, specialized for biting and exuding (top)*, brushing* (middle)*, and spraying noxious secretions* (bottom)*.*

The defenders of the termite colonies are the soldier caste. Soldiers typically have large heads, with long sharp mandibles operated by powerful muscles. Some species have specialized mandibles that can be pressed hard together.[4] When the force generated by the muscles reaches a critical point, the two mandibles snap past each other, like a finger and thumb being snapped. So much energy is released during the snap that if the mandibles strike an ant or other small insect, it can be stunned. (A similar mechanism has been independently evolved in the ant *Mystrium.*) Not all termite soldiers are specialized for offense; some, like those of the genus *Cryptotermes,* have large cylindrical heads used for temporarily plugging holes in the galleries during sieges. (Again, the same specialization has been independently evolved in several ant species.)

Many termite species use chemical warfare against their attackers, producing noxious substances in glands located in the head. The chemicals are delivered by a variety of means. Some termites exude the secretion when biting with their mandibles. Others wipe the fluid down their opponent's body with a brush formed from part of their lip. Some termite soldiers, called *nasute* in reference to nose, have pointed heads and fire their secretions through a nozzle. Although blind, they have good aim and can shoot the noxious secretions over distances of inches (several centimeters). Soldiers of the termite species *Globitermes sulphureus* secrete a yellow fluid that gels when it meets the air, entrapping the attacker, and often the termite too. The fluid, which is ejected from the mouth in large quantities, is stored in a large reservoir that extends all the way back from the head to fill a large part of the abdomen. It appears that the fluid is pressurized by the contraction of the abdominal muscles, and these contractions can sometimes be so violent that the termite's body explodes, spraying fluid everywhere! (The ant species *Camponotus saundersi* produces a similar sticky secretion, and it is also prone to explode.)

I confess that I have a problem with spiders. Not the small ones—they do not bother me at all—but those large hairy ones. How people can keep pet tarantulas and let them crawl up their arms is beyond me. An inborn wariness of spiders is natural enough, like a concern for snakes, because of their potential danger. But surprisingly few spiders are considered dangerous to humans: about twenty to thirty species out of a total of thirty thousand. A tarantula's bite, for example, is described as being barely worse than a wasp's sting. The distinction between the use of the words *bite* and *sting* is based on anatomical differences. Spiders, all of which are predatory, inject venom into their prey through a pair of hollow fangs. These are borne on the ends of biting mouth parts, called *chelicerae.* Wasps, in contrast, have a single stinger at the end of their abdomen, as do ants. The spider's fangs are mov-

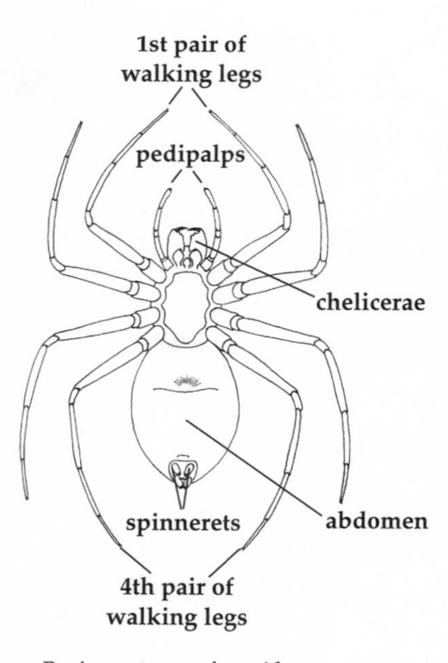

Basic anatomy of a spider.

able, and when not in use they are folded into a pair of shallow grooves, like the blades of a penknife. With the exception of a single family, all spiders are venomous.

Large hairy spiders do exist, some of which are big enough to eat small vertebrates like birds. But most spiders are small, frequently under 1/4 inch (5 mm) in length, and they feed mostly on insects and on other spiders.

Spiders and scorpions are classified together as arachnids and share a number of features, including the possession of eight legs. Insects, in contrast, have only six legs. But if you look closely at many spiders, including tarantulas, they appear to have ten legs. This is because the pair of appendages immediately behind the chelicerae (the biting parts) look like walking legs, though they are usually much shorter. These appendages, called *pedipalps*, are used for prey handling. Most spiders have eight eyes, usually arranged in two rows, one behind the other. The eyes are generally small, with limited vision, but some spiders, especially jumping spiders, have one pair of extra large eyes in front which probably form good visual images.

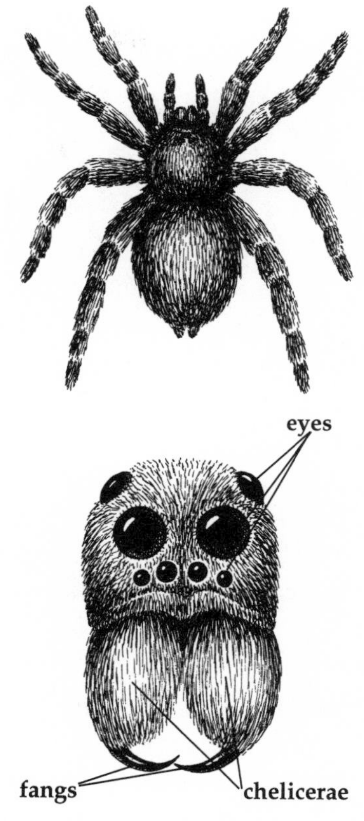

A tarantula, seen from above (top) *and head-on.*

Broadly speaking, there are two types of spiders: those that spin webs, which are used to catch prey passively, and those that actively search for prey instead of building a web. Regardless of whether they build webs, all spiders spin silk. Silk is both light and strong, having about twice the tensile strength of mild steel.[5] It is also tough, meaning that it does not break easily when exposed to moving forces. Glass and wood, for example, have similar tensile strengths and are equally strong when pulled apart. However, glass is easily broken, as when struck with a hammer, whereas wood is not and is therefore said to be tough. Spider silk, which is a polymer like nylon (a polymer is a long chain of repeated sequences of small molecules), is much tougher than man-made fibers, being almost ten times tougher than Kevlar. It is also about twenty times tougher than tendon. Recent re-

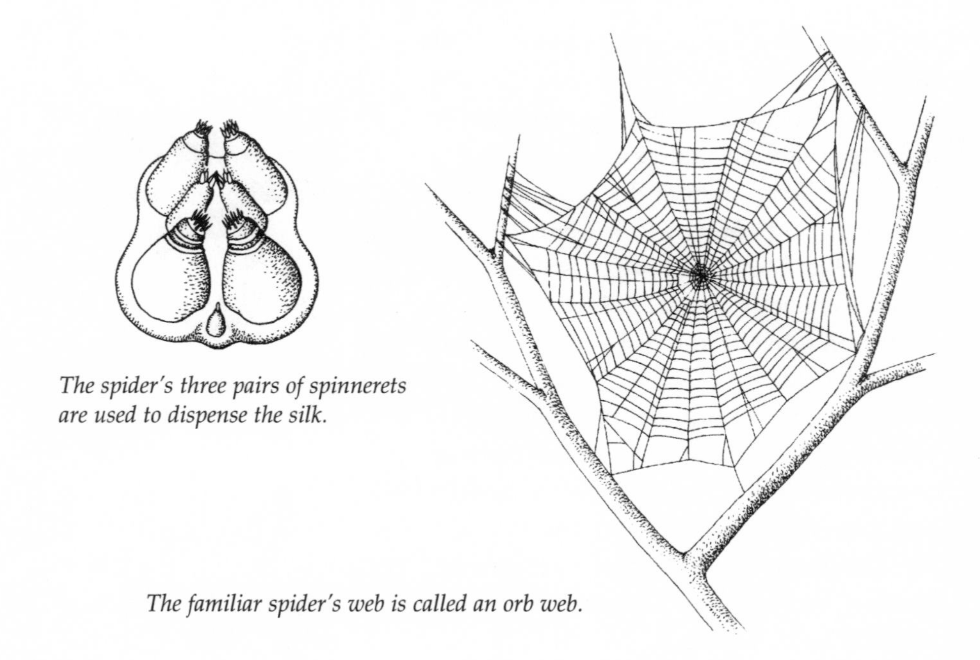

The spider's three pairs of spinnerets are used to dispense the silk.

The familiar spider's web is called an orb web.

search shows that spider silk is a composite material, made up of crystals of one amino acid (alanine) embedded in a matrix of another (glycine). Almost half of the crystals are lined up parallel with the length of the fiber, while the others seem to cross-link them with the matrix molecules, an arrangement which may explain the silk's toughness. Scientists studying spider silk are hopeful that their findings can be used to improve the toughness of synthetic fibers.

Silk is a most versatile material, which spiders use for all manner of things, from lowering themselves from great heights to tying up prey. The silk is manufactured by several different glands in the abdomen, each one producing a silk for a particular purpose. The finished products are ducted to the *spinnerets,* which are small appendages located at the end of the spider's abdomen. There are three pairs of spinnerets, which are mobile like fingers, and the silk is discharged through a number of small spigots located at their ends. The threads of silk are very fine, on the order of a few micrometers (thousandths of a millimeter), which is less than the diameter of a human red blood cell. Some threads are considerably finer, being only a fraction of a micrometer in diameter.

Spiders' webs have never been of particular interest to me. When I was a boy, I used to tweak the occasional web with a blade of grass, to watch the spider rush out to see what had just arrived for dinner, but I took no interest in the web itself. However, things changed when I started delving into the literature in preparation for writing this chapter. I discovered that the web of the common or garden spider, which I had taken for granted all these years, is an ingenious structure. Did you know that the outer zone is sticky while the central zone, the hub, where the spider often sits and waits, is not? I didn't! The stickiness helps snare any insect that blunders into the web, but how does the spider reach its prey without getting horribly mired itself? The spider can pick its way along the spokes of the web, the radial threads, which are not sticky. There is also an openly woven area just outside the hub, called the free zone, which is not sticky and where the gaps are big enough for the spider to pass from one side of the web to the other with ease. It might make perfectly good sense to us for the spider to use these nonsticky regions to avoid getting stuck. However, spider specialist Lyn Forster tells me that most of the orb spiders she has ever watched scramble all over the web, seemingly oblivious to the sticky threads. The reason they do not stick to the web may have to do with their regular grooming habit, where they draw their limbs through their mouthparts to clean them. It has been suggested that the coating of saliva the legs receive stops them from sticking to the web glue.

Being a dyed-in-the-wool empiricist, I hastened to the back garden to check these details out for myself. I expected to find lots of spiderwebs outdoors but could not find a single one. We have one or two indoors, I confide, but these are the somewhat untidy kind, referred to as *sheet webs*, not the regular-looking ones that are called *orb webs*. Where had all the spiders gone? I made the rounds again, looking in all the likely places. Then I saw a single thread in a juniper tree, glinting in the sun. Using my reading glasses, I followed the thread along and found a small web about 5 inches (13 cm) across. It was constructed of the finest silk, which explains why I had missed it the first time I had looked. And there, at the hub, was a small orange-brown spider, about $^{1}/_{8}$ inch (4 mm) long. Most web spiders are nocturnal, but this one was obviously wide awake. Using a thin blade of grass, I lightly touched one of the circumferential threads in what was meant to be the sticky zone. It was sticky. Then I checked a radial thread. It was not sticky; nor were any of the frame threads—the ones attaching the web to the tree. The spider did not remain passive during all of this. As soon as I touched the first thread, it scuttled out from the hub to see what had been caught. I tried to make my inspection as minimally disturbing as possible, but the spider, probably deciding that I was a dangerous prey item, dropped out of the web. It looked like a free fall, but the spider was probably attached to a strand of silk. This is not because the spider would hurt itself in falling a few feet to the ground, but rather to enable it to climb back up to the web once the danger had passed. Not all orb webs are the familiar rounded shape. Some, called *ladder webs*, are elongated and appear to be designed to catch specific prey items, like the ones that are "leaned up" against tree trunks for catching moths.

Some spiders are able to modify the appearance of their webs, to make them more attractive to insects. One particular species, the golden orb weaver (*Nephila clavipes*), found throughout the tropics and subtropics of the New World, can modify the color and reflectance of its silk according to local lighting conditions. In the bright light at the edge of the forest, and in clearings, the spiders spin yellow silks, giving their webs a golden hue. Foraging insects, attracted to the yellow tones commonly associated with flowers and fresh leaves, are naturally attracted to the golden webs, thereby falling prey to the spider's guile. However, in dimly lit locations beneath the forest canopy, where foraging insects appear indifferent to web colors, the spiders conserve their pigments and produce webs that look white.

Many spiders sit and wait in the middle of the web, but others use a small retreat set off to one side, made of spun silk. This is often connected to the hub by a signal thread, which the spider uses to detect vibrations when a prey strikes the web. Most of the prey are flying insects, and because of the stickiness of the web, they get themselves more firmly entangled the more they struggle. But prey do sometimes manage to free themselves, especially if they are big and strong like wasps, so the spider wastes little time in attending to them. It briefly and cautiously touches the prey with its front legs and pedipalps, keeping well clear of any dangerous parts like stings. Spiders may do this to identify what sort of prey is in the web. Not that spiders are fussy eaters—they tend to eat everything that comes their way.

Most, if not all, spiders, like ants and many other arthropods, have a keenly developed sense of smell; hence the use of pheromones in communication. They do not have olfactory organs like ours and those of other vertebrates, but have taste hairs (contact chemoreceptors) instead. These are hollow hairs that are open at the top end and well supplied with nerve cells at the other. They are especially abundant on the pedipalps and on the first pair of walking legs.

Once a web spider has decided that a new arrival is an item of food, it starts wrapping it in silk. It is usually only after the victim has been safely secured that it is bitten. However, large prey may be bitten immediately, and the biting process may last for seconds or even minutes. Spiders' venom is a mixture of neurotoxins, which attack the nervous system, and proteolytic enzymes, which break down proteins. Once the venom has done its work and the prey is subdued, the spider can begin feeding. But the carcass first has to be removed to a "safe place," either to the hub, where the web can be monitored at the same time, or to the spider's retreat. To do this the spider has to cut the body out of the web using its chelicerae and legs. If the prey is not too heavy, it is carried by the chelicerae; otherwise a short thread is attached to the food parcel, and it is hauled away using the back legs.

Spiders do not eat their prey in the normal way, by consuming solid particles. Rather, they drink their prey in a liquid form, which conjures some vivid mental images! The enzymes injected into the victim's body as part of the venom start liquefying the protein right away. This slurry, together with the victim's own body fluids, is sucked up through the puncture holes made by the fangs, using the pumping action of the stomach. At the end of the process, which may take several hours, all that is left of the insect's body is the hollow husk of its exoskeleton. Some spiders have serrated mouth parts, which they use for pulverizing the prey. Species so equipped mash up their prey, sucking and straining the pulp as they go. When they are finished, they are left with an unrecognizable mass of tissue—with bones protruding if their victim was a vertebrate.

The web takes a lot of wear and tear, not only from prey items and the inedible flotsam that becomes trapped there, but also from the wind and rain. Repairs are made as needed, and the entire web is replaced from time to time. If the catches have not been very good, the spider may decide to build a new web elsewhere. Choosing a suitable site is not random, account being taken of sun and shade (spiders like to be out of the noonday sun), prevailing air currents, and the abundance of prey. Some spiders will even set up a few test strands of silk to get an idea of prey density. It is therefore no accident that spiderwebs are frequently found close to places where there is a temporary abundance of flying insects, such as near flowers or animal droppings. Garden spiders (*Araneus diadematus*) usually replace their webs every day, though they may reuse the old supporting framework. The material of the old web is not wasted, though, because web spiders eat their old webs. Experiments with radioactively labeled webs have shown that 80–90 percent of the old material is recycled into the new web, and all within half an hour of the old one's ingestion. Other spiders, like the black widow, continue using the old web, adding to it and patching it up where necessary.

The primary objective of the spider's web is to trap other animals, and a number of different designs have evolved, some more ingenious than others. The purse-web spider (*Atypus*) spins a simple tube, closed at one end, 6–12 inches (15–30 cm) long and 1/2 inch (1 cm) in diameter. Most of the tube is buried vertically in the soil, but the rest lies above the ground. The spider lives inside the tube and when an insect crawls over the exposed section, it rushes to the attack. Biting through the web with its long chelicerae, the spider punctures the prey with its fangs. The paralyzed victim is then dragged through the hole made by the bite. Once the web has been repaired, the spider begins feeding.

Some spiders build webs that act like spring traps. In one of these types the spider (*Hyptiotes*) builds a triangular web which is anchored at two corners. The third corner is attached to a thread that the spider hangs on to, thereby keeping the web held open. As soon as a victim strikes the web, the spider lets go of the thread, collapsing the structure all about the prey, thus ensuring its entrapment. But the prize for ingenuity must go to spiders like *Steatoda castanea,* which build a snare called a *frame web.* Such a web is a loose affair of mooring lines, horizontal supporting threads, and vertical trap threads. The trap threads, which are beaded with sticky droplets, are under much tension and are easily broken. When an insect blunders into one of the trap threads, it snaps, whisking the victim into the air. The prey, which is often an ant, struggles to get free, striking other trap threads which also break, securing it more tightly. The resident spider, alerted by all of the vibrations, quickly climbs down to the victim, adding more sticky threads before delivering the coup de grâce.

While the web builders are probably most familiar to us, the majority of spiders do not build webs. Most of them live on the ground, taking shelter in the forest litter or in burrows in the soil. Included among these wandering spiders are the wolf spiders, a large family (Lycosidae) of some 2,500 species. The group name suggests they pursue their prey like wolves, but they are essentially sit-and-wait predators. Some of the wolf spiders of the Mediterranean region of Europe are referred to as *tarantulas,* named after the Italian town of Taranto. However, the term is more widely used for large tropical spiders, belonging to an entirely different group (the mygalomorphs). Many wolf spiders are less than 1/2 inch (1 cm) long, but the large ones reach body lengths in excess of 1 inch (2.5 cm). This may not sound very large, but when their long robust legs are taken into account, they are large enough to cover your hand. Wolf spiders are nocturnal. Their days are spent beneath the ground in silk-lined burrows, which they leave at night to search for prey. They have one prominent pair of eyes among the other six, and although they can perceive movements, they do not appear to be particularly good at forming images. They can probably see things only when they are up close—inches rather than feet. Like most other spiders, they are very sensitive to vibrations, both airborne ones and those transmitted through the ground. The airborne vibrations, which seem to be especially important to wandering spiders for locating prey, are detected by specialized sensory hairs called *trichobothria*. These fine hairs are attached to a thin membranous disc, which is supplied by several nerve endings. The slightest air movement, including sound waves and the wing beats of flying insects, causes the hair to move, vibrating the disc and setting off nerve impulses. These "touch-at-a-distance" receptors are arranged in regular rows on the legs and enable the spider to get an exact fix on its prey. Having located a suitable target, the spider stalks toward it, and when within grasping range, seizes it with the tips of its front legs. The prey is pulled toward the chelicerae and bitten, then held away from the body by the chelicerae while the venom takes effect. This minimizes the risk of damage to the predator and prevents the prey from getting a grip on the ground. Spiders often fasten their prey down with silk threads prior to feeding, especially if the victim is large, a strategy particularly important for those few spiders that live in trees.

The trap-door spiders (family Ctenizidae) build an underground lair, the entrance to which is guarded by a hinged trapdoor. The inside of the door is lined with silk, like the walls of the burrow, but the outside incorporates soil and plant litter so that it blends in perfectly with the rest of the terrain. When a suitable prey animal strays too close to the camouflaged door, the spider, vigilantly waiting on the other side, rushes out and seizes it. Trap-door spiders are often large, and capture a wide range of prey, including small vertebrates, which are usually smaller than themselves. Indeed, most of the predators treated in this chapter attack prey that are smaller than themselves, in contrast to the situation we saw among mammalian predators. One notable exception, the cobweb spiders (which build irregular-looking webs), can trap and kill prey up to three times larger than themselves. Certain other spiders can also tackle larger prey; so too can some centipedes, scorpions, and "false spiders" (more about them later). The one feature these arthropods have in common is that they are all venomous.

A jumping spider.

I was sitting in a friend's back garden in New Zealand, on a bright Sunday afternoon, when I noticed a small spider crawling along the leg of the picnic table. There was nothing unusual in this, but the next instant the spider sprang from the table and landed on the ground, a few inches below. This was my first encounter with a jumping spider. The jumping spiders (family Salticidae) belong to a large group, numbering about four thousand species, that hunt by jumping on their prey. Unlike many other spiders, they are diurnal, being active during the day. They are all small, less than $^{1}/_{2}$ inch long (1 cm), and their most prominent feature is a pair of enormous eyes that occupies most of the "facial" region. These anterior eyes have a very narrow field of view and are of a fixed focal length, with no means of changing their focus. They face directly forward and are able to move from side to side and up and down, in tandem. Because their fields of view do not overlap, they do not give any binocular vision. However, when a potential prey is sighted, the eye movements become very rapid, and their two fields of view then overlap with those of the next pair of eyes, giving the spider depth perception. The anterior eyes have a high degree of visual acuity, providing the spider with a clear visual image of the prey when it is directly in front of it. Their maximum range appears to be about 8 inches (20 cm). The next pair of eyes lie on either side of the anterior ones, and these also face forward. They have a much wider field of view, of about 55°, and since their two fields overlap, they afford binocular vision, therefore depth perception, especially when overlapping with the anterior eyes. The function of the remaining two pairs of eyes is concerned primarily with detecting movement. One of the pairs, positioned behind the large pair, face outward. They too are large and have a wide field of view of 130°, enabling the spider to detect movements on a wide arc around the body.

The first indication that a jumping spider has caught sight of a potential prey is revealed by a slight tensing of its legs and body. The spider then swivels its body around to face its prey. When its body is aligned and the forward-facing eyes have locked onto the target, the spider moves forward, either at a run, or a walk, or sometimes a slow stalk. Once the distance has been closed to an inch or so (a few centimeters), the spider pounces on its victim, driving its fangs into its back. Jumping spiders are so fast that they can lock onto the flight trajectory of a fly, pouncing for the kill as soon as the insect lands. Some jumping spiders prey on web spiders by jumping onto their webs and attacking them. Other species are more subtle in their approach and enjoy a higher rate of success by disguising their true identity. Instead of jumping onto the web, which would immediately put the resident spider on the defensive, they either walk onto the web or lower themselves on a strand of silk. Before the resident realizes that this is not another web spider, the invader pounces and kills it.

Even more dastardly than this tactic is the deceit practiced by *Ero*, a spider that often targets a particular species of orb-web spider called *Meta segmentata*. The deceiver approaches the web of a female spider and plucks at a special thread that is used by males when they come calling. The female, thinking that a potential mate has arrived on the doorstep, leaves the hub of the web to investigate, only to be attacked and eaten. *Ero*, which also attacks other spiders, belongs to a group called pirate spiders (Mimetidae), which prey exclusively on spiders. Their strategy is to bite their victim on the leg, their fast-acting venom causing paralysis within seconds. Another group (the Zodariidae) preys only on ants, similarly attacking with a bite on the leg. *Callilepis nocturna*, an ant hunter belonging to a different group, uses a frontal approach, aiming its bite at the base of the ant's antenna. The attacker then withdraws to a safe distance from the enraged ant, which could inflict a lethal wound with its powerful mandibles, and waits. Within a few seconds the injured antenna wilts, and the ant starts walking around in circles, the direction depending on which antenna was injured. The stricken ant is easily relocated, and after making sure it is paralyzed, the spider carries it off. Finding a safe haven, the spider builds a temporary shelter of silk and begins feeding.

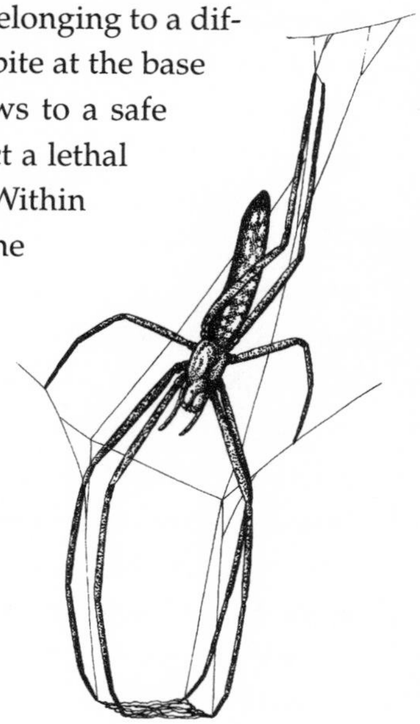

The ogre-faced spider (*Dinopis*), named for its two enormous eyes, weaves a small rectangular web between its front legs, like a cat's cradle. It then hangs upside down from a thread and waits for a suitable prey to pass by. These spiders hunt mostly at night, their huge eyes giving them good night vision. When a prey is spotted, it is scooped up in the web,

The ogre-faced spider hangs upside down with a small web stretched between its front legs, waiting to pounce down on a passerby.

additional threads being spun around it to completely entangle it. The prey is bitten one or more times during capture, and feeding quickly follows.

In chapter 2 we saw how some snakes spit venom, and there are some parallels in the spitting spider *Scytodes.* The venom gland of this spider is huge and occupies much of the anterior body segment. The front portion of the gland secretes toxins, while the rest produces a sticky material like glue. The gland is well supplied with muscle fibers, whose contractions cause the fluid to be squirted forcibly out of the spider in two streams. When a suitable prey passes within range—within about 1 inch (2.5 cm)—it is sprayed with two zigzag streams of fluid, gluing it to the ground and paralyzing it at the same time.

The most sophisticated weapon system of any spider has to be that of the bolas spider (*Mastophora*). As the name suggests, this spider captures its prey by striking it with a sticky droplet that it throws at the end of a thread. Even more remarkable, the bolas spider launches its attacks at night, and against a flying target. Its prey are certain kinds of male moths (noctuid moths like *Spodoptera*), which it attracts by releasing a scent into the air that mimics the sex pheromone of the female moth. While the amorous male circles overhead, the spider takes aim and flings its bolas. This technique is not always successful, but a spider is often able to catch two or three moths in an evening. If the spider is that successful, it will probably not hunt again for many more days, because spiders have low metabolic rates and correspondingly low food requirements.

The sit-and-wait hunting strategy that typifies many spiders often leads to long periods without food. Spiders have been known to live for over six months without eating. When prey become available, though, they make the most of it and can consume more than half of their body weight in one meal. Part of the reason they can do this is that their abdomens, unlike those of most other arthropods, are not encased in a rigid exoskeleton and are expandable. Some spiders have short lives, lasting only one year, but others, including the large tarantula types, live for twenty years or more. It is usually only the females that live long lives; the males, which are often smaller than the females, have much shorter life expectancies.

Almost all spiders are solitary and appear to be intolerant of other spiders' company, which is why people who collect spiders never put more than one specimen into a container. Their mutual intolerance may help explain why the diminutive males of some species approach the females with some caution. The black widow spider (*Latrodectus mactans*), whose bite can be fatal to humans, has the reputation that the female eats her mate after copulation. However, the reputation appears to be largely undeserved because the male usually withdraws from the female without coming to any harm. But there are some species where postcoital cannibalism of the male is the rule. In the garden spider (*Araneus pallidus*) the male, which is much smaller than the female, appears to be incapable of copulating without constantly sliding off her abdomen. The only solution appears to be for the female to bite into his abdomen to keep him in place. But within a few minutes of doing this she starts eating him.

Male spiders are generally regarded as the unfortunate victims of their larger and more aggressive partners, attacked at a time when they are least able to defend themselves. However, some recent studies on the Australian redback spider (*Latrodectus hasselti*), close relative of the black widow spider, have shown that the male of the species is a willing participant in these acts of cannibalism. And for his own benefit! It is difficult to imagine how being eaten could possibly benefit the individual, but from the perspective of the male's evolutionary fitness—his ability to pass his genes on to the next generation—his demise is in his best interests. Females lay egg sacs containing several hundred eggs, and there is much competition between rival males to fertilize them. Each female mates with upward of three males, and the duration of their copulation, which she appears to control, is from about six to thirty minutes. Sperm transfer continues for the duration, so if the copulation time can be extended, the male is able to deliver more of his sperms, thereby fertilizing more of the eggs. During the act of copulation the male does a somersault in which he places his abdomen directly over the female's mouthparts. If she decides to eat him, she does so while he is in this position. Significantly, copulation lasts about twice as long if the female is engaged in devouring her mate than if he escapes with his life. Furthermore, she is unlikely to mate with another male after having consumed a previous suitor. Since males appear able to mate only once in their life, they have to make the very most of the occasion. Their lives are short too—only a few months compared with up to two years for females[6]—so their suicidal act of procreation is not such a big sacrifice. Incidentally, the tips of the male's paired copulatory organs, which are modified pedipalps, break off inside the female during copulation, which may explain why males apparently mate only once.

Whatever negative thoughts we may harbor about spiders, they are very beneficial to us humans because they destroy so many insects that feed on our crops. Good gardeners know better than to kill a spider, and the same is true for certain farmers. Chinese farmers have been using spiders to control pests for generations. The spiders are housed in bamboo tepees, which are moved, en masse, to fields in need of pest control. The employment of spiders has reduced the use of chemical pesticides in China by a staggering 60 percent. Unfortunately, though, the practice is being abandoned in favor of less labor-intensive westernized farming methods. Ironically, there is now a movement in the United States to use spiders to control insects, as in soybean fields.

I always thought wasps were only interested in eating fruit and nectar and other sweet things—they only become a nuisance in our garden when there are autumn apples on the ground. While wasps generally are vegetarian, there are times of the year when they become predatory. To understand why requires some knowledge of their life history.

Some wasps and hornets are solitary, but many are colonial, living in communal nests like honeybees. Colonial species have a caste system, like ants, with workers leaving the nest each day to forage for food. They build their nests from wood pulp, constructing a series of adjoining cylindrical cells in which the larvae will develop. One of the commodities the workers have to gather is therefore rotten wood, which they chew into a pulp by mixing it with saliva. Water also has to be collected and is used during hot weather for cooling the nest. The workers visit flowers to collect nectar for food, but when the larvae start getting bigger they demand flesh.

Hornets and wasps are aggressive predators, usually searching for prey while on the wing. They hunt during the day, including butterflies, bees, and flies among their quarry. A frequently used strategy is to strike at their prey while they are feeding on flowers, knocking them to the ground during the attack. They kill by biting with their mandibles, often decapitating their victims in the process. Their stinger, which is located at the end of the abdomen, is seldom used. The location of prey appears to be done visually, and wasps will apparently even fly through open windows to attack flies they have seen on the other side of the glass. Wasps of the *Vespula* genus have been observed to attack butterflies in flight, taking them to

Wasps sometimes attack butterflies on the wing.

the ground, where the wings are snipped off using the mandibles. While this may serve to immobilize the captive, the removal of portions of the body, like the head, is just part of the usual preparation for transporting the muscle-filled thorax back to the colony. Hornets and wasps often attack bees, tearing open their abdomens to feed on the honey-filled crop. In some parts of the world the predation on honeybees poses a serious problem for beekeepers. Lone wasps (*Vespula orientalis*) have been seen approaching a hive and then retreating, thereby enticing one of the guards to pursue it. Once the honeybee is far enough away from its hive, the hornet turns and pounces on it. Sometimes several hornets will collaborate to attack a hive.

Wasps sometimes attack birds. In Britain, workers of the wasp species *Dolichovespula sylvestris* were seen buzzing around a pair of blackcaps (*Sylvia atricapilla*) that were sitting on their nest. The parents were eventually forced to abandon their nest, which the wasps immediately plundered. They killed the three newly hatched fledglings, tearing up their bodies with their mandibles and transporting the flesh back to the colony. Wasps can carry loads weighing up to half their own body mass. If the prey is too large to be carried in one trip, they break the carcass up into pieces, taking as many trips as necessary to ferry it back.

Many solitary wasps lay their eggs underground, usually laying several eggs in each cell. But before sealing the cell, the female provisions it with food for the larvae that will hatch from the eggs. If she used fresh meat, it would probably go bad by the time the eggs hatched (two to four days). She therefore captures live caterpillars, stinging them with sufficient venom to subdue them and arrest any further development, but without killing them. The female lays an egg; then a caterpillar, now in a state of suspended animation, is butted and pushed into the narrow cell. Once safely stowed on top of the wasp's egg, she lays another egg and repeats the process. Once the cell has the full complement of eggs and caterpillars, the female seals the opening. When the wasp larvae hatch, they remain underground for about a week, feeding on the living caterpillars and growing strong.

The saga of hunting and killing among the arthropods has no end, but I will draw it to a close with a brief account of scorpions.

There is nothing very sophisticated about scorpions. No webs, no snares, no subterfuge. They just seize their prey with their pincers (pedipalps) and rip them apart with their chelicerae. If the prey are large or dangerous, though, they are first immobilized with the stinger at the end of the "tail." Scorpion venom, which is usually neurotoxic, differs from that of

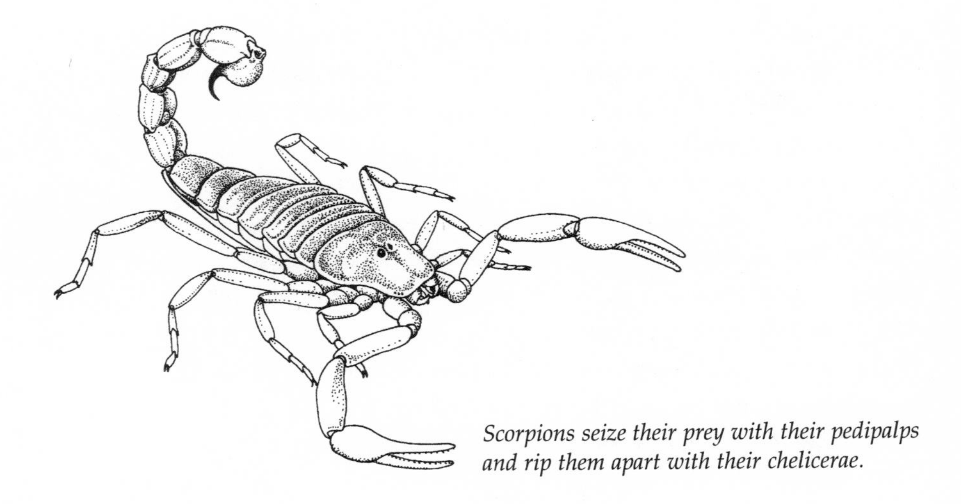

Scorpions seize their prey with their pedipalps and rip them apart with their chelicerae.

most spiders and snakes in that it usually lacks enzymes (there are a few exceptions) and therefore plays no role in the digestive process. Scorpion venom is also unusual for having multiple toxins, some of which are particularly toxic to certain types of animals. The venom of a given species may be especially effective against insects, while another may target mammals. In spite of their sting, most scorpions are harmless to humans and only about 25 of the 1,500 species are potentially fatal. Scorpions account for an estimated five thousand deaths each year, placing them third behind snakes and bees. Some species are small, less than 1/2 inch (1 cm) long, but most scorpions are the size of a mouse rather than a fly. The largest ones, like the African species *Hadogenes troglodytes,* reach lengths of over 8 inches (20 cm). Most are nocturnal, but they can readily be studied at night by using an ultraviolet lamp ("black light"), which causes their bodies to fluoresce. Like spiders, they are primarily sit-and-wait predators, feeding on most of the animals that happen to come their way. Their prey include spiders, centipedes, other scorpions, large insects, even snakes, lizards, frogs, and rodents. They have well-developed eyes, which are among the most light sensitive of all eyes, but most locate their prey by detecting air movements, using the same specialized sensory hairs (trichobothria) as used by spiders. They are generally not very active but can move very fast when they want to, and can snatch flying insects out of the air. Many species spend over 90 percent of their lives in relative inactivity in their underground burrows. This is because their metabolic rates are so low that they do not have to feed very often. In this regard they compare with spiders, and scorpions have been known to survive for more than a year without food. Their frugal demands on the environment make them ideally suited to deserts, but they do occur in other environments too, even on the seashore. *Vaejovis littoralis,* for example, lives along the Gulf coast of California, feeding on crustaceans (*Ligia*) and other intertidal animals.

Most scorpions live in warmer climates, but a large colony has become established in England, on the site of an old dockyard on the Isle of Sheppey, about 30 miles (50 km) outside London. They have been there since the late nineteenth century, introduced from overseas by ships' cargoes. The species *Euscorpius flavicaudis* is common throughout western Europe, but this colony is one of the most northerly populations of scorpions in existence.

In some locations scorpions are the top predators and can have a detrimental effect on some of the species upon which they prey. This has been shown in experiments where scorpions of a particular species have been removed from an area, resulting in a dramatic increase in their prey species. However, since scorpions have such low metabolic rates, only a fraction of the total population of a given species will be feeding on any one night. So although their biomass may be greater than that of many other species, their prey consumption relative to their own biomass is likely to be lower than that of other predators. Scorpions are also eaten by other species, both invertebrate and vertebrate. Among the former are wolf spiders and black widows, centipedes, and solpugids (Solifugae).

Solpugids, sometimes called "false spiders" or "wind scorpions," look like spiders but lack the spider's narrow waist. They have huge chelicerae, more hair than spiders, and are often larger and are more aggressive. In short, they are a spider-fearing person's worst nightmare! When a group of us from our museum were digging for dinosaurs in the Chihuahuan desert a few years ago, there was a torrential rainstorm that turned dried gullies into raging rivers. Desert storms cause temporary blooms in insect populations, and this, in turn, brings out the arachnids in force. Our campsite was crawling with horrors for several days after the rains, including the most fearsome looking solpugids. Fortunately for me, I was on a supply run in town and missed the worst of it.

The solpugids that invaded our camp no doubt took a heavy toll on the scorpion population. But the greatest invertebrate threat to scorpions probably comes from others of their kind, both of the same, and of other, species. In almost all attacks between scorpions, it is the larger individual that wins. Antagonists face each other, locking their pincers on their opponent's appendages and reaching over the top with their stingers. Since the larger individual has the longer reach, it is more likely to deliver the first sting. Some species are immune to their own venom, but they can still kill one another by delivering a sting into one of the abdominal nerve centers. Large individuals achieve this end by turning a smaller adversary onto its back and attacking the vulnerable areas. Large species prey heavily upon smaller species, just as large individuals eat large numbers of their smaller kin. The North American scorpion *Paruroctonus mesaensis,* for example, is a large species and is responsible for about half of the predation on other scorpion species in its own area.

The female of the species is generally larger than the male, as in spiders, and many females kill and eat their partner after copulating. In many species, including *P. mesaensis,* the females outnumber the males, largely due to the females' penchant for cannibalism.

Among vertebrate predators, birds take the largest numbers of scorpions, accounting for 37 percent of all vertebrate predation, followed by lizards (34 percent), mammals (18 percent), frogs and toads (6 percent), and snakes (5 percent). Owls, being nocturnal, target foraging scorpions, while other vertebrates dig them out of their burrows during the daytime. Some predators are immune to scorpion venom, such as certain snakes (*Chionactis occipitalis*) and the mongoose (*Herpestes edwardsi*). Others, like baboons (*Papio*), meerkats (*Suricata*), and the grasshopper mouse (*Onychomys*), simply remove the scorpion's stinger before it can harm them, by breaking off the tail.

It is probably the scorpion's lack of sophistication and meager demands on the environment that explain its long tenure on Earth. Scorpions appeared on land over 300 million years ago, which is at least 50 million years before the Age of Reptiles. And what a remarkable age that must have been to have lived through.

7

Saurian Warriors

A dense white fog swirls across the land, enveloping all but the treetops in its humid embrace. They float above the ephemeral sea, rose pink in the dawn light, disembodied from reality. The air is pleasantly warm, but when the sun drives off the mist it will become oppressively hot. A lush plain will then be revealed, stretching as far as the eye can see and dotted with swamps and shallow ponds. What a contrast to the dry season, when the swamps dry out and the ponds revert to parched clay pans.

The Jurassic flora was dominated by ferns and conifers, many of which were similar to modern forms. But the giant tree ferns and the cycads—primitive plants that look like palm trees growing out of pineapples—would look unfamiliar to most of us. Insects were the most abundant invertebrate animals, as they are today, and although mammals were present, the dominant land vertebrates were the reptiles. Reptiles flew in the air (pterosaurs) and swam in the sea (ichthyosaurs, plesiosaurs, and others), while the dominant terrestrial reptiles were the dinosaurs. It is because reptiles were the dominant vertebrate group that the Mesozoic Era, comprising the Triassic, Jurassic, and Cretaceous periods, is often called the Age of Reptiles. The Mesozoic began about 245 million years ago, ending 180 million years later, 65 million years ago.

A gray shape materializes out of the mist and takes on a more solid form. As it moves closer, it resolves itself into a small dinosaur. At 10 feet (3 m), it is a little

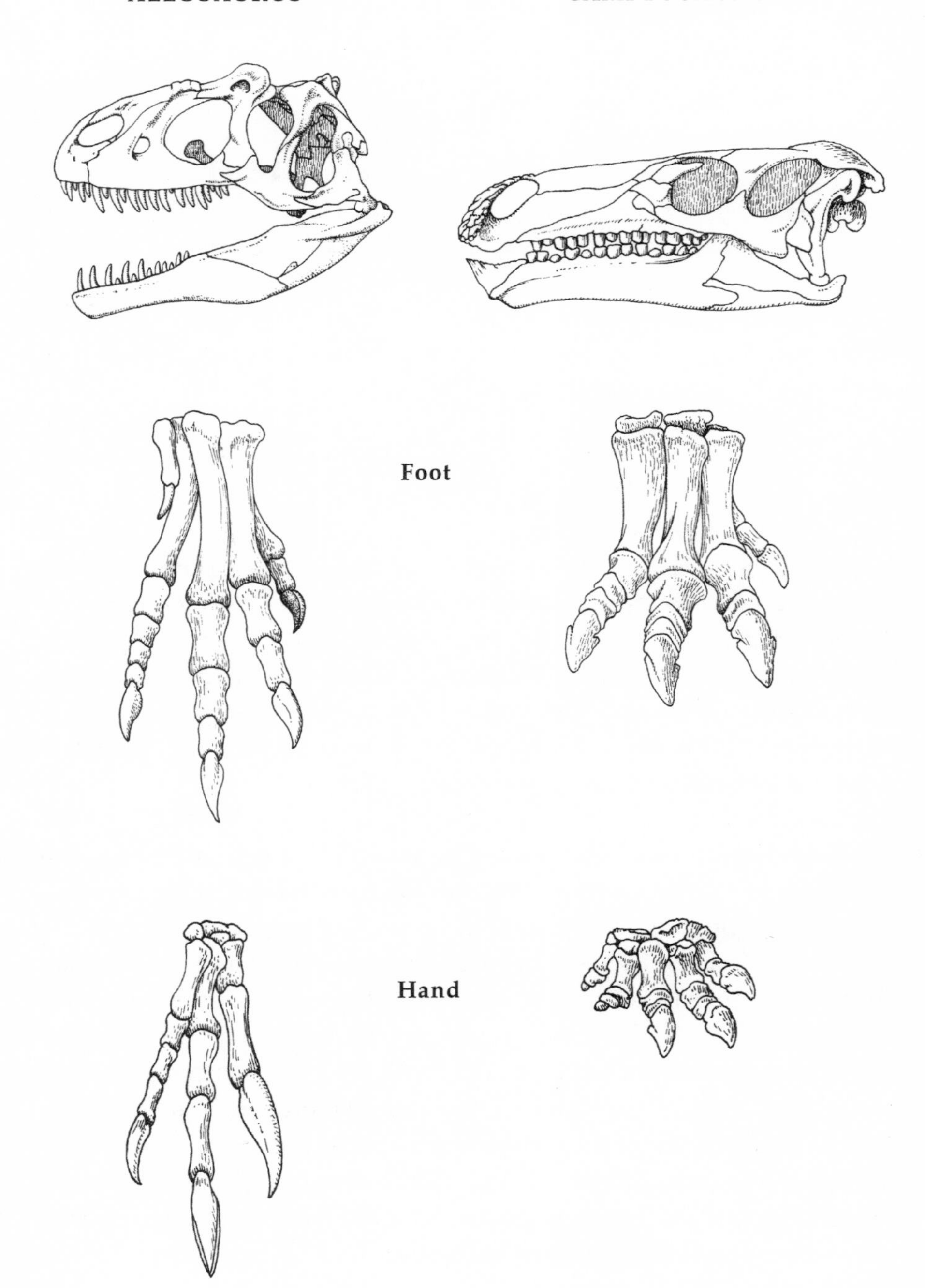

A comparison of the skulls, feet, and hands of Allosaurus *and* Camptosaurus *shows some obvious adaptations to their respective meat-eating and plant-eating modes of life.*

longer than a cow, but half of its length is occupied by a gently tapering tail, so it weighs much less. The dinosaur, an inoffensive plant-eater called Camptosaurus, *moves unhurriedly through a clump of tree ferns. It is an immature individual and has much more growing to do—adults reach lengths up to 23 feet (7 m). It walks semierect, on its hind legs, using its muscular tail to counterbalance its fore part. It is an amply built animal, a bulldog rather than a greyhound, with thick columnar legs, and three-toed feet ending in stout claws. The forelimbs are short and robust, and the hands, looking reminiscent of our own, have five stubby fingers, each terminating in a blunt claw.*

The tree ferns all look alike to our human eyes, but the camptosaur selects one from among the others and begins feeding. The fronds at the top of the feather-duster tree are more tender than the rest, and the dinosaur stretches up to reach them. Since the dinosaur is immature, it is not yet wise to the ways of the world. Had it been more wary, it might have detected a skulking presence in the blanketing mist and foliage.

Using both hands, it pulls the choice fern fronds down toward its jaws. A thick black tongue slithers out of its mouth and wraps around the green plant—a python seizing its prey. The tongue is retracted, pulling the living plant into its mouth and the mechanical chewing jaws. Another frond catches its eye, and as it strains up higher to reach it, the entire tree explodes in a shower of leaves. The dinosaur is thrown to the ground with a bruising thud. But before it has a chance to stagger to its feet, a massive set of jaws clamps around its neck. The yard-long skull bears down with an inexorable force, shearing through flesh and bone like a guillotine. The attacker, a 39-foot-long (12 m) Allosaurus, *shakes its short powerful neck so vigorously that the* Camptosaurus *is lifted from the ground, breaking its neck and almost severing its head. Warm blood pumps from the lacerated vessels in the neck, spraying the green ferns red.*

The allosaur has not eaten for almost two weeks and immediately sets to work on the carcass, still twitching with involuntary muscle spasms. Unlike its prey, Allosaurus *is not a fastidious eater, devouring flesh, skin, and gut with indifference. Its voracity borders on malice.*

How realistic is this depiction of death in the Jurassic? Were the large meat-eating dinosaurs[1] like *Allosaurus*—sometimes called carnosaurs—active predators? Perhaps they were merely scavengers, as some paleontologists believe. How do we know they were flesh-eaters anyway? What are the differences between herbivores and carnivores? And were dinosaurs really warm-blooded, like us, or cold-blooded like a snake, relying on the sun for warmth? Before trying to answer these questions, I must emphasize that we have no way of knowing, for sure, what life was like during the Mesozoic. It was, after all, a very long time ago. All we can do is draw inferences or make educated guesses, based upon the fossils themselves. There are some paleontologists, though, who would have us believe we know everything there is to know about dinosaurs—from their metabolic rates to how fast they ran—but this is wishful thinking. Just consider the evidence available to paleontologists. At the very best we have a complete skeleton to study, but even this is only part of the animal. We have none of the soft tissues, though we may sometimes have clues to these in the form of skin impressions, muscle insertion scars on bones, endocranial casts of the brain, and the like. We cannot watch the animal move, observe its behavior, measure its temperature, record its heart rate, or measure any of its other body functions. We seldom know its sex, we can rarely determine how old it was when it died, and we cannot always be sure whether it was an adult or a juvenile. If we have several skeletons that look the same, we can probably never be sure whether they belonged to the same species, because, with rare exceptions, we have no way of recognizing a biological species in the fossil record. (A biological [animal] species is a natural entity in the living world and is recognized by the fact that its members freely interbreed.) The best safeguard against overinterpreting fossil animals is to make constant reference to present-day ones. If it is not possible to draw conclusions regarding a particular feature from the skeleton of a modern animal, it will not be possible from a fossil one either. Paleontologists have no more knowledge of the extinct beasts they study than neontologists (those who study present-day animals) would have of modern animals if their conclusions were based solely on skeletal remains. Imagine if elephants had not yet been seen in the flesh. A neontologist would need a very vivid imagination to invent the elephant's trunk, because it is formed entirely of soft tissue. And who would surmise from examining the skeleton of a hippopotamus that these animals are amphibious rather than terrestrial?

Most of the clues to an animal's diet are to be found in its skull, the teeth being especially informative. Paleontologists usually have no more difficulty in distinguishing between carnivorous and herbivorous dinosaurs than neontologists do in distinguishing between lions and antelopes. *Allosaurus* has long, sharp, daggerlike teeth, whose edges are serrated like a steak knife. Their efficiency in puncturing and slicing through flesh is in no doubt. The teeth are set in deep sockets, with about as much of the tooth embedded in bone as there is exposed. As a consequence, upper and lower jaws are deep. The meat-eaters had to have their teeth firmly rooted to prevent them from being pulled out during struggles with their prey. Teeth were lost from time to time, though, and so were continually being replaced, as in modern reptiles. Alternate teeth were replaced, a strategy for preventing adjacent teeth from being replaced at the same time, which would have resulted in gaps.

Camptosaurus, in contrast, has small, spatulate teeth, suitable for cropping plants, and these teeth are not rooted in deep sockets. There are also differences in the structure of the jaw joint, which determined the extent of jaw mobility. As in modern carnivores, the jaw joint of meat-eating dinosaurs allowed for only an up-and-down movement of the jaws, with no lateral or fore-and-aft mobility. In modern carnivores this is necessary for the efficient slicing action of the teeth, just as a pair of scissors with a tight screw is better at cutting than one with a loose joint. Herbivores, in contrast, have a loose jaw joint, enabling the jaw to be rolled from side to side to grind up the plants prior to swallowing.

The head of *Allosaurus* was supported on a short and seemingly powerful neck. It is assumed that the neck was powerful, that is, muscular, because the long neural spines and ribs associated with the cervical (neck) vertebrae provided large attachment areas for the neck muscles, which were, by implication, large. This contrasts with the cervical condition in *Camptosaurus,* where the neural spines are barely developed and the cervical ribs are relatively small. The deep skull and lower jaw of *Allosaurus* also provided a large attachment area for the jaw muscles, giving a powerful bite. *Camptosaurus,* in contrast, has a narrower skull and lower jaw, modified for chewing rather than biting. Modern big cats similarly have well-developed neck and jaw muscles, allowing for the powerful movements of the head and jaws during prey capture and killing (see chapter 1). The suggestion that *Allosaurus,* and other large carnivores, were scavengers, feeding only on carrion, rather than active predators, is difficult to reconcile with some of their skeletal features. Why would a scavenger need such a large and powerful head and neck for feeding on putrefying meat. Why have such formidable teeth and such a powerful bite? And what is the relevance of *Allosaurus* having sharp claws on the hands and feet if not for capturing living prey?

Whether predatory dinosaurs chased down their prey, like mammalian carnivores, or laid in ambush, like crocodiles and most other modern reptiles, is not so readily answered. The question hinges on dinosaurian metabolism and body temperatures, a highly speculative subject in which opinions vary widely among paleontologists. Robert Bakker stands at one extreme of the spectrum, with his view that all dinosaurs had high metabolic rates and high body temperatures, like modern birds and mammals. Others, myself included, are of the opinion that most dinosaurs probably had much lower metabolic rates than comparably sized birds and mammals, more like those of modern reptiles. I would need an entire chapter to discuss the subject,[2] so will limit myself to a few key points. First, dinosaurs were quite unlike any animals alive today, and we should not expect them to have conformed to any physiological patterns that are familiar to us now. The possible exception to this are the smallest and seemingly the most active dinosaurs, like the dromaeosaurs (more on them later),[3] whose metabolic rates and physiologies may have been comparable to those of modern birds. Second, there is no reason to believe that all dinosaurs had similar metabolic rates and thermal strategies, any more than do all mammals today. There are, for example, mammals like sloths that have low metabolic rates, and others, like bats and bears, that lower their body temperatures during roosting and during hibernation.

We saw in chapter 2 that large reptiles, like the Komodo dragon, tend to retain body heat, just as hot potatoes keep warm longer than hot peas. This phenomenon is described as *thermal inertia.* The giant tortoises of the Galápagos Islands weigh up to about 450 pounds (200 kg), giving them such a large thermal inertia that their body temperatures fall only about 3 C° (5 F°) at night, although ground temperatures fall by as much as 20 C° (36 F°). Their body temperatures therefore remain fairly constant throughout the twenty-four-hour period.[4] Most dinosaurs were huge compared with Galápagos tortoises, so a constant body temperature would have been an inescapable consequence of their massive thermal inertias. Many dinosaurs, including many of the meat-eaters, possess skeletal features, such as their long lower leg segments, indicative of their running abilities. Much heat is generated by muscles during running, and since this tends to be retained by large animals, because of their large thermal inertia, the body temperatures of the more active dinosaurs were likely to have been quite high. Indeed, getting rid of excess body heat may have been a problem for some dinosaurs, especially the largest ones. The problem would have been exacerbated if their metabolic rates had been high, like those of modern birds and mammals. High metabolic rates are also costly in terms of food consumption. So why should dinosaurs have evolved high metabolic rates if they were already enjoying the benefits of constant and fairly high body temperatures by virtue of their large thermal inertias?

To summarize, it seems likely that dinosaurs, like modern mammals and birds, had various thermal strategies and different metabolic rates. It also seems likely that their metabolic rates were lower than those of birds and mammals, probably more like those of modern reptiles. If most dinosaurs did indeed have relatively low metabolic rates, how might this have affected their activity levels and the ability of the predatory ones to hunt their prey?

Modern reptiles lack stamina (see chapter 2), and their bursts of activity during prey capture, as when a Nile crocodile captures and drowns a wildebeest, are powered by anaerobic metabolism. Animals cannot keep up anaerobic exercise for very long, largely because of the accumulation of lactic acid in their muscles. Reptiles have difficulties breaking down lactic acid, which is why they become exhausted after prolonged bursts of strenuous activity. Crocodiles, which belong to the same group as dinosaurs (the Archosauria), and which are their closest living reptilian relatives,[5] have a high tolerance to lactic acid. Furthermore, large individuals are able to keep up strenuous exercise for longer periods than are small ones. Crocodiles weighing more than about 200 pounds (90 kg), for example, take over half an hour of strenuous exercise to become exhausted. Most dinosaurs were much larger than this, and if they had metabolic rates like crocodiles, and similar abilities to tolerate lactic acid, they would have been capable of maintaining long periods of rapid activity before becoming exhausted. However, dinosaurs may have had a more efficient respiratory system than crocodiles, and this may have enabled them to be more efficient at breaking down lactic acid. So even if dinosaurs had low metabolic rates like those of modern reptiles, they would still have been able to maintain high activity levels for prolonged periods. And there is no need to suppose that all dinosaurs had the same, low, metabolic rates as crocodiles. It is therefore conceivable that some predatory dinosaurs were capable of pursuing their prey, like modern mammalian carnivores, rather than being restricted to laying in ambush, as do most modern reptiles. What sort of prey would the predators have attacked? Would an *Allosaurus* have been capable of bringing down a large sauropod?

The sauropods, those quintessential dinosaurs with long necks and long tails, like *Diplodocus* and *Brachiosaurus,* were the largest animals ever to walk the Earth. Finding adequate words to convey the enormity of these animals is not easy. The only way you can get a feel for their immense size is to stand beside a skeleton in a museum. *Brachiosaurus,* one of the larger sauropods, but by no means the largest, is the biggest skeleton that has

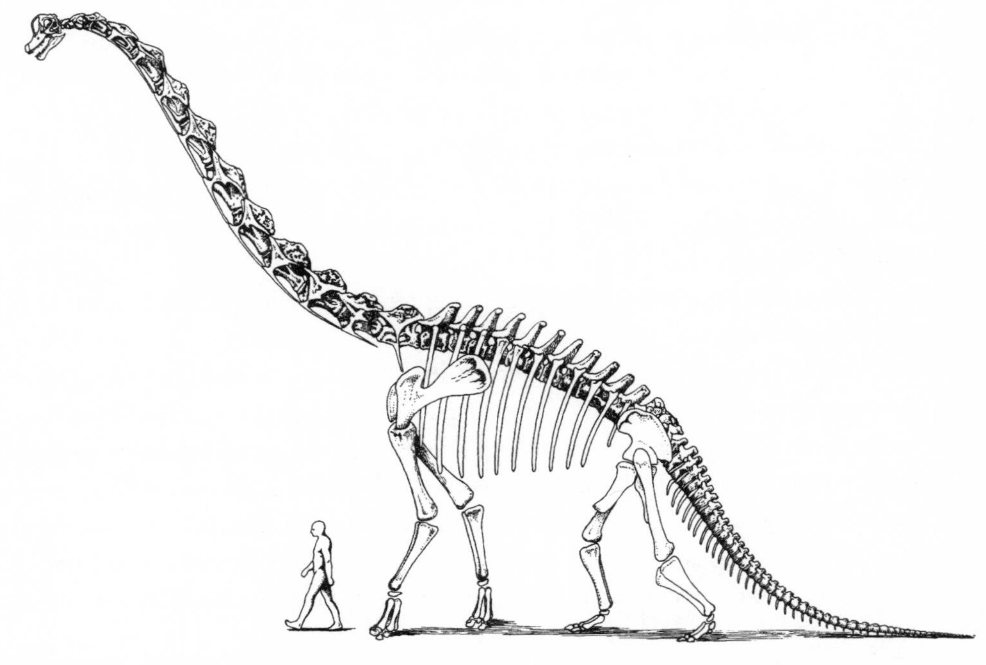

Brachiosaurus, *the largest dinosaur skeleton ever mounted, dwarfs a human figure.*

ever been mounted. The specimen is on public display in Berlin, and I have been fortunate enough to visit the museum and handle some of its bones.

If you have ever stood on the Tarmac beside an airliner, you will appreciate what it feels like to be dwarfed. The airliner analogy is not inappropriate because it is estimated that, in life, *Brachiosaurus* would have weighed as much as a Boeing 727—about 78 tons.[6] The Berlin specimen is mounted on a waist-high plinth, making direct comparisons of height more difficult, but if I were standing on the same level as the skeleton, my head would be on a level with its elbows. And I estimate that if I walked beneath it, I would be unable to reach up and touch the ends of its ribs. As with a giraffe, the forelegs are longer than the hind ones, raising the shoulders above the hips so that the entire vertical column is inclined upward. The head therefore towers so high above the ground that it almost touches the ceiling and can barely be seen. This is one of the reasons that a plaster skull has been mounted on the skeleton and the original skull is on exhibit in a display case.[7] The skull is surprisingly small for the size of the skeleton—about as big as a horse's.

While being overwhelmed by the enormity of *Brachiosaurus,* I glanced across the gallery at what appeared to be the skeleton of a juvenile sauropod. It looked so small in comparison, but it turned out to be an adult *Diplodocus.* As it happens, the *Diplodocus* skeleton is a little longer than that of *Brachiosaurus,* 85 feet (26 m) compared with 75 feet (23 m), but it looks so much smaller because *Diplodocus* is more slenderly built. It also has a relatively longer tail than *Brachiosaurus* and is nowhere near as tall. *Diplodocus* had an estimated live weight of 18 tons, which is about three times heavier than an African elephant. *Apatosaurus,* formerly known as *Brontosaurus,* was marginally shorter than *Diplodocus*—75 feet (23 m)—and is now known to have been of similar build, so the previous weight estimate of 28 tons is probably too high. The massive size of sauropods convinced earlier paleontologists that they must have lived in lakes and ponds, where their bodies would have been buoyed up by the water. However, there is compelling evidence, mostly from sauropod trackways, that they did walk on land, though that does not mean that they may not have also spent some of their time in water, as do modern elephants, given the chance.

Given the large weight disparity between *Allosaurus,* which weighed only about 2 tons, and its contemporaneous sauropods, what are the chances that it might have preyed upon them? Although most modern predators tend to prey on animals smaller than themselves, thereby avoiding injury and possible death, there are some exceptions. The Komodo dragon, for example, is a lone hunter that sometimes attacks and kills water buffalo weighing as much as ten times its own weight. *Allosaurus* appears to have been no less formidable, and lone individuals may have successfully preyed upon the lighter sauropods, like *Diplodocus* (18 tons). There is even some circumstantial evidence from dinosaur trackways to suggest that sauropods may have been stalked by predatory dinosaurs. The trackways in question, from Texas, comprise twenty-three individual sets of sauropod footprints, all pointing in the same direction and apparently made on the same occasion. They bear testimony to the passage of a herd of great dinosaurs, over 150 million years ago, perhaps along an access route beside a river or a lake. And along with the sauropod tracks were the prints of large carnivores. Some of these had been imprinted by the sauropod tracks, showing that the carnivores had passed along that way before the herd of sauropods. Did they stop farther along the way and wait in ambush for the sauropods? Other carnivore tracks crossed over the sauropod prints and had clearly been made by predators that had traveled the route after the passage of the sauropods. In one case the track of a carnivore was imprinted alongside that of a sauropod. When the sauropod swung to the left, so did the carnivore. Was the predator following the sauropod? Perhaps the herd was being stalked, and there may have been other predators waiting in ambush for them farther along the way. But it is also possible that the carnivores were merely traveling along the same route.

If sauropods were attacked by predators, either by lone hunters or by hunting packs, what defensive strategies might they have deployed? Some paleontologists think that sauropods may have used their long tails like bullwhips, an idea that has always struck me as somewhat fanciful. Even more imaginative is the suggestion that sauropods reared up high on their hind legs to fight off their attackers with their forelimbs. Aside from the difficulty I have in visualizing how an animal weighing several times more than an elephant could be so nimble, I can't reconcile the additional burden placed on its vascular system by raising its head so far above the level of its heart. Perhaps sauropods, like elephants, relied exclusively on their large size and herding instinct to discourage predators. This would probably work well against lone hunters, but if a pack of carnivores attacked a lone sauropod or managed to separate one from a herd, especially a young one—like wild dogs attacking a herd of wildebeest—they would probably succeed. The sight of an *Allosaurus* pack attacking a giant sauropod like *Apatosaurus* would probably be every bit as protracted and gory as a pod of killer whales attacking a blue whale.

The Cretaceous period was even more spectacular than the Jurassic from the perspective of predatory dinosaurs. *Tyrannosaurus,* one of the largest terrestrial killers ever to stalk the Earth, lived during this period.[8] Imagine a 7.5-ton carnivore with a head almost as long as your body that was tall enough to peer into a second-story window. This is the stuff of horror movies, but this 46-foot-long (14 m) predator did exist. Like its Jurassic predecessors, *Tyrannosaurus* and its Cretaceous relatives (collectively called tyrannosaurs) were slenderly built, bipedal dinosaurs, with long serrated teeth, and sharp claws on the hands and feet. In striking contrast to Jurassic genera, the forelimbs were diminutive, having been reduced from three fingers to two. The forelimbs were so small compared with the rest of the skeleton that they would have appeared abnormal, like birth defects. Since they were barely long enough to reach the mouth, even with the head bent down, it is difficult to visualize their function. One suggestion is that the fingers may have been used as grapples to secure the front end of the body when a recumbent individual returned to its feet. Given that elephants seldom lay down, and then only for brief periods, because of the risks of compressive tissue damage, the heavier *Tyrannosaurus* may not have laid down at all. Perhaps the forelimbs were used to grapple with prey. Alternatively, they may have been important when the individuals were young, when the forelimbs may have been much longer relative to the rest of the body. The phenomenon where parts of the body grow at a different rate than the rest is common, and is called allometry (meaning different length). The head, for example, grows more slowly that the rest of the body, which is why babies have disproportionately large heads compared with adults.

Hadrosaurs were the most likely prey species of the tyrannosaurs. Not only were they among the most abundant herbivorous dinosaurs of the time, but they had no armor or weapons of defense. Hadrosaurs, or duck-billed dinosaurs, were about the same size as the Indian elephant, having an estimated body weight of 4 tons and averaging about 25 feet (7.8 m) in length. There were many different genera and species, and many of them had elaborate crests on their heads. Some, like *Corythosaurus,* had rounded crests, flattened from side to side like a cock's comb. The most extensive crest was that of *Parasaurolophus,* which had a long curving tube projecting from the back of its head that was over 3 feet (1 m) long. Most of the crests, including those of *Corythosaurus* and *Parasaurolophus,* were hollow and formed part of a complex series of ducts and hollow chambers conveying air from the nostrils to the back of the throat. There has been much speculation on their possible function, the most likely explanation being that the crests served for species recognition and communication. The hollow crests are suggestive of resonating chambers, and it seems very likely that they were used to produce loud sounds. From their size and configuration, they would have produced low-frequency sounds, like those of elephants, and

Horned dinosaurs, like Triceratops, *bear an obvious resemblance to living rhinos.*

these would have traveled over long distances—elephant sounds travel for about 2 miles (3 km). There is evidence that hadrosaurs lived in herds, and being able to communicate by low-frequency sounds would have enhanced their ability to warn each other of approaching danger.

Hadrosaurs could only defend themselves against predators passively, through herding and flight, but some of their herbivorous contemporaries possessed an impressive armory of defensive structures. Most of the horned, or *ceratopsian,* dinosaurs, for example, had formidable horns. They also had an extensive bony shield extending from the back of the skull that would have protected the neck region. *Triceratops,* one of the best-known ceratopsians, was a massive animal, about 30 feet (9 m) long and weighing an estimated 6 tons. Its superficial resemblance to a living rhino is quite apparent, and like the rhino, it probably would have been a dangerous animal to attack. Whether *Triceratops* were capable of charging at their adversaries, as often depicted in illustrations, is debatable, though. None of their skeletal features are suggestive of great speed; quite the reverse, but they may have been able to make short lunges. Like hadrosaurs, they appear to have lived in herds, and they might have formed defensive circles—horns facing out—when under attack.

The ankylosaurs were the dinosaurian equivalent of tanks. They were broad and squat, with a thick but flexible armor protecting the back. This armor consisted of a series of bands of bony plates and nodules embedded in the skin. The top of the skull was also protected with bony plates, making ankylosaurs effectively impenetrable from above—some even had bony armor in their eyelids. The shortness of their legs gave them a low center of gravity, which would have made it very difficult for a predator to flip them over onto their backs to expose their unprotected underside. Some of them may even have squatted down when faced with danger, making them almost invulnerable to serious damage. Many of them, including *Euoplocephalus,* one of the better-known genera, had a bony tail club, which was probably used for striking at their attackers. Since their bodies were so close to the ground, swinging the club from side to side would have given them a good chance of striking their attackers on the ankle or shin. Not only would this have been very painful, but it could well have broken bones. Our museum—the Royal Ontario Museum—has a juvenile individual of *Albertosaurus*—close relative of *Tyrannosaurus*—which may have received such an injury. An unusual feature of the skeleton, which is only about 20 feet (6 m) long, is that the fibula (the bone that runs parallel to the shin) was broken sometime during life. This break is revealed by a large bony overgrowth, midway along its length. It is conceivable that the injury was received when the young, and perhaps inexperienced, predator attacked an ankylosaur, but this is entirely speculative. *Euoplocephalus* was about 20 feet (6 m) long, but the largest ankylosaurs reached lengths of about 33 feet (10 m). Such large and heavily armored dinosaurs would probably have been left unmolested by the predators of the day.

The vision of a 7-ton tyrannosaur mauling a living dinosaur with its massive jaws is sufficiently disturbing, but the thought of a swarm of waist-high predators slashing and ripping an animal to shreds with their scalpel-sharp claws is chilling. These small and lightly built dinosaurs, collectively called dromaeosaurs, ranged in size from the 6-foot-long (1.8 m) *Velociraptor* and *Dromaeosaurus,* with bodies not much bigger than geese, to the 10-foot-long (3 m) *Deinonychus,* the best-known genus, which was about human sized. These remarkable little dinosaurs were discovered in Alberta in 1914, but the original material was fragmentary, and they remained poorly understood for half a century. Then, in 1964, a rich new fossil locality was found in Montana. During the next few years excavations at the site, led by dinosaur specialist John Ostrom of Yale University, unearthed a number of well-preserved skeletons, some of them almost complete. They all belonged to the same type, for which Ostrom coined the generic name *Deinonychus,* meaning "terrible claw." This name was chosen for a remarkable feature of the animal: a large, scimitar-shaped claw on the inside toe of each foot. What is more, the claw was retractile, like a cat's claw, enabling it to be kept clear of the ground during walking and running, thereby keeping it razor sharp. Many more unusual features were discovered when the rest of the skeleton was examined, some having to do with the workings of the foot. The femur has an extra process at its upper end, believed to be for the attachment of a muscle specialized for kicking. The pelvis is also modified in that the anterior prong, the pubis, which is usually directed forward in carnivorous dinosaurs, appears to point backward. This modification was presumably to allow for a wide swinging arc of the hind leg. Lastly, the tail, aside from a short section nearest the pelvis, is ramrod stiff and straight. The stiffness is achieved by an overlapping series of long bony processes extending from each vertebra. This sort of tail, while unusual, is not unique to the dromaeosaurs, and a somewhat similar tail occurs in the ornithomimid, or ostrich-mimic dinosaurs.

The forelimbs of *Deinonychus,* in marked contrast to those of *Tyrannosaurus* and its relatives, are very well developed, suggestive of strength,

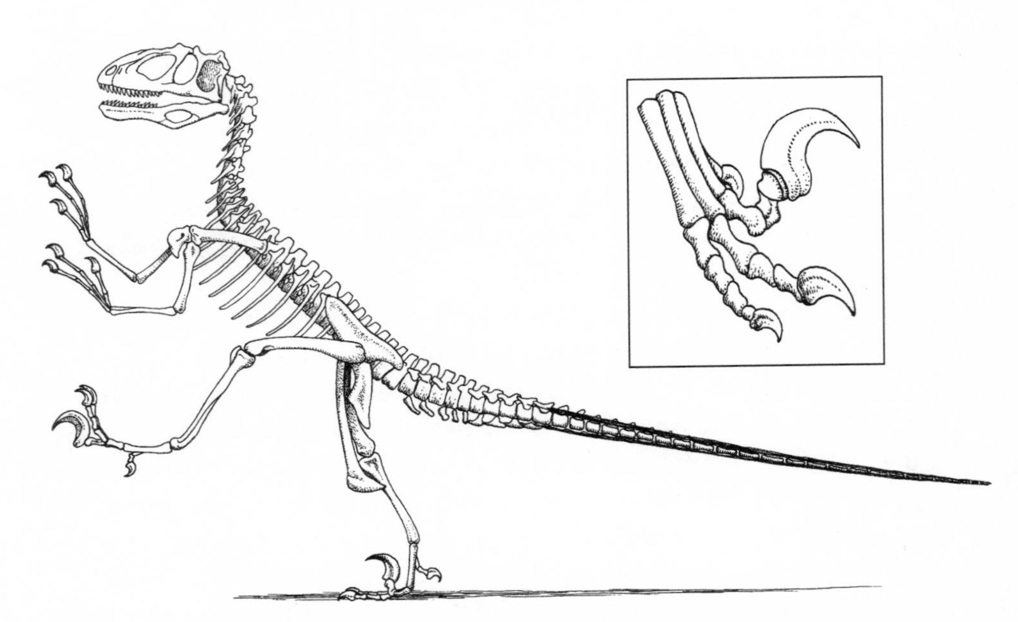

Deinonychus, *a small, lightly built dinosaur, was named for the long retractile claw on its foot* (inset).

with three long fingers, each ending in a sharp talon. The structure of the wrist and forearm shows that the hands were capable of rotation, conferring a grasping function reminiscent of that seen in cats. Like the rest of the skeleton, the skull is lightly built, with a slender lower jaw. A study of the muscle insertion scars on the lower jaws—these appear as roughened areas, ridges, and depressions—points to a rapid and powerful snapping action. The teeth are sharp and serrated, like those of other meat-eating dinosaurs, but are unusual in being markedly sloped toward the rear.

What conclusion are we to draw from the skeletal specializations of these remarkable dinosaurs? How did they capture and kill their prey? The idea that dromaeosaurs—the "raptors" of the book and film *Jurassic Park*—hunted in packs, like African hunting dogs, is an attractive one. This is partly suggested by their small size, their estimated body weight ranging from 65 to 180 pounds (30–80 kg). There is also evidence of pack hunting in the fossil record. At one of the sites where *Deinonychus* was excavated, the remains of at least four individuals were found in association with fragments of a single and large individual of the herbivorous dinosaur *Tenontosaurus.* It appears that the predators had attacked the herbivore, then all of them had perished in some catastrophe—perhaps a collapsed riverbank. An even more remarkable find was unearthed in the Gobi desert in 1971 when a complete skeleton of *Velociraptor* was found, embracing the skeleton of the small-horned dinosaur *Protoceratops.* British paleontologist David Unwin's study of the specimen shows that they were locked together in mortal combat when they were buried by a sandstorm. The *Velociraptor* holds the head of *Protoceratops* with the sharp claws of its fingers, while the other bites one of its attacker's arms in its beaky jaws. Their death struggle has been frozen in time for over 80 million years.

Given that dromaeosaurs probably hunted in packs, how did they capture and kill their prey? From their leg proportions, and the lightness of their skeleton, they were probably fleet-footed and agile, and likely quite capable of leaping from the ground. The teeth, being curved and backwardly inclined, were probably more suitable for seizing and tearing than for killing. The forelimbs were also specialized for seizing, the sharp talons acting as grappling hooks. Killing would probably have been accomplished by the formidable talon of each foot. Using its stiff tail for balance, the dromaeosaur may have stood on one leg and slashed at its prey with the other. Alternatively, it may have leaped from the ground and slashed with both feet together. The slashes may have been directed at the vulnerable underbelly, to disembowel the unfortunate prey. They may also have been used to inflict deep wounds elsewhere, inducing extensive blood losses. It does not require a vivid imagination to visualize the destructive powers possessed by these reptilian assassins.

A late Cretaceous afternoon. A herd of hadrosaurs grazes in a clearing at the edge of a forest of bald cypress and gum trees. They munch unhurriedly on the ground cover, vacant eyes staring into vacant space. One of their number stops feeding and lifts its head. It remains in its four-footed feeding posture—short forelegs propped against the ground, head held alert—for several long moments. Has it heard something? The other hadrosaurs show no response. The moment passes, and the hadrosaur begins feeding again.

Something small, fast, and unseen by hadrosaurian eyes slips from behind a gum tree and streaks toward a cypress stump, several yards closer to the herd. The diminutive creature, a blur of emerald and saffron, disappears behind the stump as if it never existed. It is not alone. Others of its kind are advancing in similar fashion toward the edge of the forest, like a platoon of marines.

The hadrosaurs, now aware of the danger, have started running about in all directions, front legs tucked tightly against their bodies, tails held stiffly out behind. One of the hadrosaurs, seized by blind panic, breaks away from the others and races off toward the trees, less than one hundred yards away. The very next instant the dromaeosaurs break cover, running like fiends to narrow the gap. Realizing its mistake too late, the hadrosaur tries to swerve around, but the nearest attacker has already launched itself at its quarry's neck. The dromaeosaur misses its mark but manages to bite a loose fold of skin behind the hadrosaur's arm. Razor-sharp teeth and claws dig in deeply, attaching the diminutive dinosaur to its prey like a leech. It inflicts minimal damage, but the shock of the attack startles the hadrosaur, giving a second assassin the opportunity to launch an attack. A hind leg is seized and badly lacerated. Unfortunately for the dromaeosaur, it is dashed to the ground the next instant, crushing the life from its frail body. But its work is done because the hadrosaur falters on its next step, allowing a third dromaeosaur to join the fray. Minutes later the hadrosaur falls to the ground beneath a biting, slashing fury of dromaeosaurs.

Some of the attackers that have encircled the doomed hadrosaur balance on one leg while slashing with the other. Others leap into the air and slash with both talons. Many more have swarmed on top of the hadrosaur. Using their front claws as grapples to steady themselves, they slash down, first with one leg then with the other. Rivulets of blood trickle from countless wounds, turning the hadrosaur's gray hide red. The mournful bellows seem to go on forever, but, with one final grunt, the hadrosaur finally dies. Its body continues twitching for several more minutes, by which time most of its attackers have satiated their cravings for warm flesh.

Thomas Hawkins, that eccentric fossilist of the early 1800s who sold his collection of extinct reptiles to the British Museum amid a great furor (it was discovered, after the sale, that some of the skeletons had been "improved" upon by the addition of missing parts), published two large tomes on extinct reptiles. In one of these was a fanciful reconstruction of life in the remote past. In a dark and brooding seascape, two plesiosaurs, locked in a death struggle with an ichthyosaur, writhe and thrash in the background. A dead ichthyosaur lies beached on the rocks in the foreground, two ghoulish pterosaurs in attendance. One of the pterosaurs appears to be stabbing at its eyes, while the other gazes toothily at the mortal combat being fought in the sea. Hawkins was right to depict the pterosaurs beside the ocean because almost all specimens have been collected from marine sediments, and most species are considered to have lived in close association with the sea. But their predilection for carrion was pure speculation on Hawkins's part, and most of them appear to have eaten fish. Their fish-eating habit hardly fits into our current theme of reptilian warriors, but they had some interesting specializations for catching their prey, and there are certain parallels with some of the animals we have already discussed, so we will spend a little time with them.

Pterosaurs, like bats, had membranous wings formed of skin, which appears to have been fairly thin. However, instead of being supported by most of the hand, it was supported only along its leading edge. The pterosaur's thumb, index, and middle fingers are short and end in hooked claws, but the fourth finger is greatly elongated and supported about half the entire wing. A remarkable feature of pterosaurs is the extreme thinness of their hollow bones, making their skeletons very light. This, coupled with the thinness of their wing membranes, would have given them very low wing loadings. We have seen that bats have lower wing loadings than comparably sized birds (see chapter 5), contributing to their enhanced maneuverability. Pterosaurs are believed to have been exceedingly maneuverable too.

There were two types of pterosaurs, long-tailed and short-tailed. The long-tailed ones were the earliest, appearing about 220 million years ago, during the Triassic period. But they became extinct toward the end of the Jurassic, at about the time when the short-tailed ones first appeared. Many of the early members of each type were fairly small, about the size of pigeons, but some of the Cretaceous ones, which were all of the short-tailed kind, were huge. *Rhamphorhynchus* was a long-tailed genus from the Juras-

The long-tailed pterosaur Rhamphorhynchus *was a common Jurassic genus. The sharply pointed teeth face forward, making them suitable for impaling fishes.*

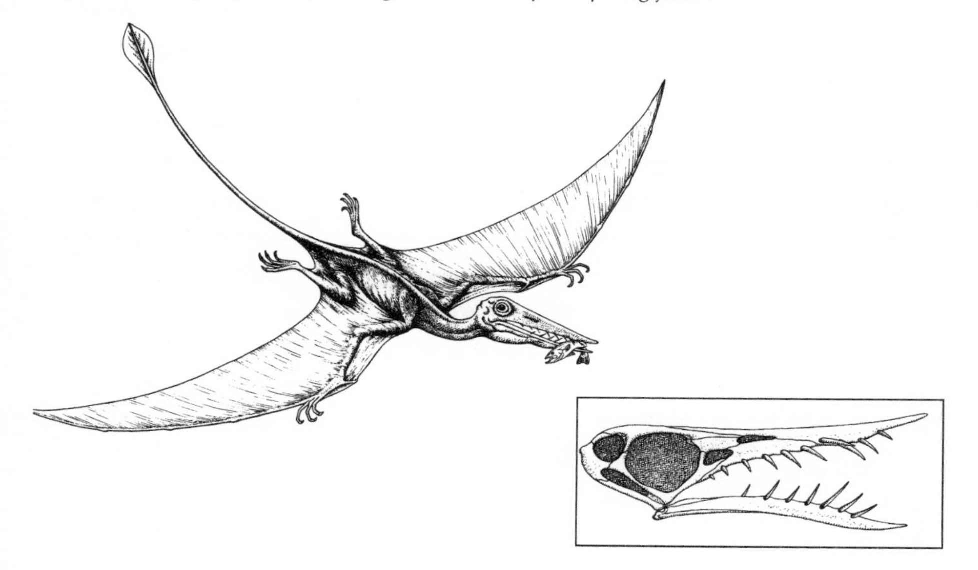

sic, primarily of Europe. The smallest species had wingspans of about 16 inches (40 cm), but the largest ones reached spans of almost 6 feet (1.8 m). The long straight tail was stiffened by overlapping bony extensions of the vertebrae, as in the dromaeosaurs, and ended in a vertical membranous flap, which may have functioned as a rudder during flight. The teeth are sharply pointed and sloped forward, making them ideally suited for impaling fish. There is direct evidence supporting a fish-eating diet because the remains of a small fish were found in the stomach region of one of the specimens. The tip of the lower jaw is sharply pointed and was probably invested in a long pointed bill. In one species (*Rhamphorhynchus longicaudus*), the tip of the lower jaw was flattened from side to side, suggesting that it may have fished like a skimmer. The skimmer bird, whose lower bill is longer than the upper one, catches fish by flying along just above the surface, with its lower bill skimming through the water.

Fish remains have also been found in the stomach region of *Pteranodon*, one of the largest and best-known pterosaurs, from the Cretaceous of Kansas. The largest species of this short-tailed genus had a wingspan of about 30 feet (9 m). Unlike most other pterosaurs, *Pteranodon* was toothless and had a large crest projecting from the back of its skull. There is much variation in the shape and size of the crest; some individuals, the minority, have a large crest, while the others have a relatively small one. Significantly, the individuals with large crests are, on average, bigger than the others, by about 50 percent. If a bar chart were plotted of numbers of individuals against body size, there would be two peaks. A somewhat similar bimodal distribution would be obtained for adults of our own species. Here the first peak, corresponding to the smaller individuals, would represent females, the other peak the males. Since the average man is much less than 50 percent larger than the average woman, the two peaks would be far less distinct. Although the bimodal distribution of body size and crest size in *Pteranodon* is suggestive of sexual dimorphism, other explanations are possible. Significantly, though, the smaller individuals had a different-shaped pelvis, with a much larger

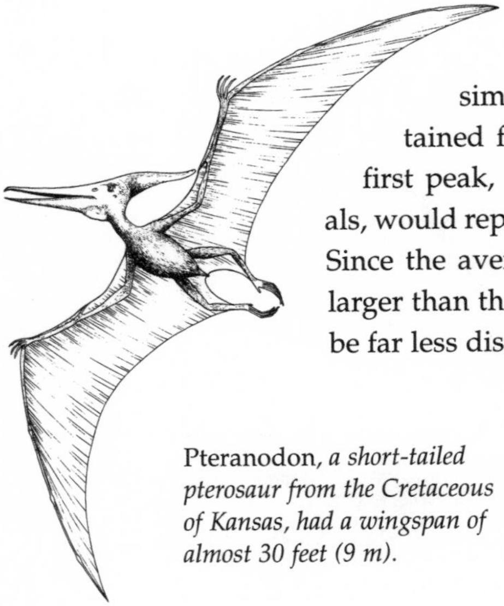

Pteranodon, *a short-tailed pterosaur from the Cretaceous of Kansas, had a wingspan of almost 30 feet (9 m).*

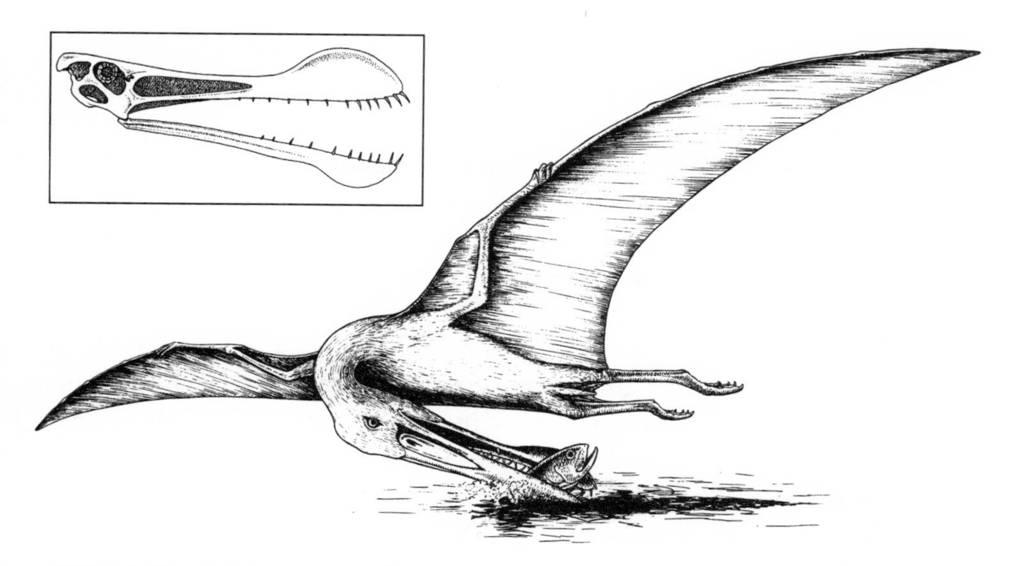

Tropeognathus, *from the Cretaceous of Brazil, has a keel on the tip of its snout and another on its lower jaw. These may have functioned as stabilizers while it fished on the wing.*

pelvic canal, appropriate for the passage of eggs. It therefore seems likely that the smaller individuals with the small crests were females, while the large-crested ones were males. This suggests that large crests may have evolved in *Pteranodon* for sexual display.

Paleontologists have suggested a way in which *Pteranodon* may have been able to fish while on the wing, without losing much forward momentum. The pterosaur would soar through the air, just above the surface, and when it spotted a suitable fish it would rotate its head down and back on its flexible neck. The fish would be seized while the head was still rotating to face toward the feet, thereby minimizing the loss of forward momentum. The head would then be flicked out of the water. *Pteranodon* may have had a throat pouch like a pelican's; remains of fish and crustaceans were found just beneath the lower jaw in one specimen, suggestive of such a structure, but the evidence is tenuous.

Tropeognathus, from the Cretaceous of Brazil, was somewhat smaller than *Pteranodon,* with a wingspan of 20 feet (6 m). The remarkable feature of this genus is the thin bony keel on the tip of its snout, with a matching one on its lower jaw. It is thought that *Tropeognathus* may have fished on the wing, as suggested for *Pteranodon,* and that the keels may have acted as stabilizers for the head as it was skimmed through the water. The keels could only have functioned in this way while the head was pointing backward; otherwise they would have had the opposite effect and caused the head to be jerked to one side. Alternatively, the keels may have been used during sexual displays.

Not all pterosaurs were fish-eaters. Some, like *Ctenochasma,* from the Jurassic of Europe, appear to have fed on small animals that they filtered from the water. The filtering apparatus was formed by numerous bristle-like teeth—over 250 of them. The animal would have waded in the shallows, scooping up mouthfuls of water, which it strained through its teeth. There were several other filter-feeding pterosaurs, but *Pterodaustro,* from the Cretaceous of Argentina, was the most remarkable of them all. This medium-sized pterosaur—average wingspan 4 feet (1.3 m)—had a long and slender skull, about 9 inches (23 cm) long, with upturned jaws. The upper jaw had numerous short and blunt teeth, but the lower jaw was unique for its dense array of long slender teeth, reminiscent of a baleen whale's bristles. Indeed, they were once thought to be bristles, but electron microscopy of some well-preserved material from Brazil shows that they have enamel and dentine, just like regular teeth.

The largest pterosaur was discovered in Texas during the early 1970s, in nonmarine sediments, which is unusual. Estimates for the wingspan vary, but somewhere between 36 and 39 feet (11–12 m) is probably close, which is about the size of a light twin-engined airplane. Named *Quetzalcoatlus* after the Aztec deity, this huge pterosaur was the largest animal ever to have flown. Like *Pteranodon,* it was toothless, but differed in having an unusually long neck. There has been much speculation on what this remarkable animal did for a living. One suggestion is that it was a carrion feeder, using its long neck to probe deeply inside the carcasses of dead dinosaurs, like some reptilian vulture. The giant carcass might also have provided a suitable launching pad for takeoff, too, but I have difficulty imagining a flesh-gorged *Quetzalcoatlus* becoming airborne in such a fashion. Another suggestion is that

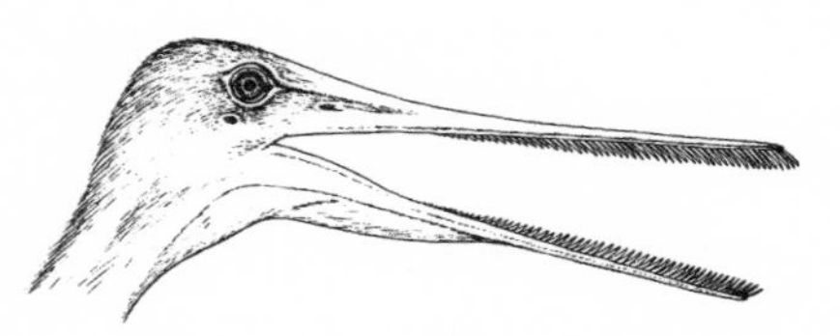

Ctenochasma, *from the Jurassic of Europe, had numerous bristlelike teeth which probably served as food strainers.*

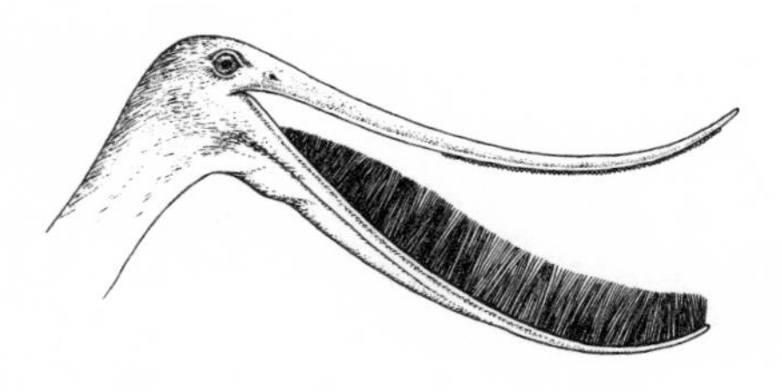

Pterodaustro, *from the Cretaceous of Argentina, had the most remarkably long filter-feeding teeth, once thought to be bristles.*

Quetzalcoatlus may have been like a giant heron, plucking fishes from freshwater with its long bill. I have no better suggestion to offer for its possible lifestyle, so perhaps this is my cue to turn to more familiar ground.

The ichthyosaurs are a group of marine reptiles contemporaneous with dinosaurs. However, they became extinct earlier, slipping into oblivion about 25 million years before the end of the Cretaceous. Dinosaurs are so popular today that it is doubtful whether anyone has never heard of them, but the same is hardly true of ichthyosaurs. However, the popularities of the two groups were reversed during most of the last century, ichthyosaurs enjoying the approbation of genteel Victorian society. The reason for this was that ichthyosaurs were discovered before dinosaurs, and their remains were far more complete than those of the early dinosaur finds.

The discovery of the first ichthyosaur, the oft-told story of Mary Anning, has all the elements of a Dickens novel. Mary Anning was born in 1799, to a humble cabinetmaker of Lyme Regis, on the south coast of England. There is a story of a great thunderstorm, and of a bolt of lightning that struck and killed the young woman who was looking after her, together with two fifteen-year-old girls. The baby Mary was miraculously unharmed, but the experience was said to have destined her for greatness. Her father, Richard, used to augment his earnings by selling the fossils he collected along the foreshore. He was often accompanied by the young Mary and her older brother, Joseph. He taught his children how and where to look for fossils, and his daughter proved to be a particularly talented pupil. His wife, also named Mary, was a fossilist as well, but this fact is seldom mentioned. The Jurassic limestones and shales that outcrop along this section of the Dorset coast are rich in fossils, and west-country folk had scoured the beaches and searched the cliffs for generations. The curiosities they found included "devil's toenails" (the oyster, *Gryphaea*), "thunderbolts" (the bullet-shaped internal skeletons of the squidlike belemnites), and "verteberries" (the vertebrae of ichthyosaurs). Lyme Regis was a seaport, as well as a thriving seaside resort for the affluent few who could indulge in such luxuries. The fossils were sold to the visitors as keepsakes of their holidays.

Richard Anning did not enjoy good health, and he died in 1810. His widow and two children were entirely dependent upon the meager earnings of Joseph, who worked in a local furniture shop. Economic necessity therefore forced Mary Anning to become a professional fossilist at the tender age of eleven.

Her first major find was apparently in 1811, when she unearthed a

4-foot-long (1.3 m) skull of some unknown beast,[9] together with parts of its shoulder girdle and some vertebrae. The specimen appears to have been discovered and partly collected by her brother, at the end of the previous year, but the exact date of the discovery is uncertain. The young fossilist probably enlisted the help of local quarrymen to excavate the specimen, and she subsequently sold it to Henry Host Henley, upon whose land it had been found. He paid twenty-three pounds for the specimen, a large sum in those days and probably a small fortune to the Annings. Henley later sold or donated the sea dragon's petrified remains to William Bullock, entrepreneur and owner of the Egyptian Temple, a private museum in London's fashionable Piccadilly. The entire collection came under the auctioneer's hammer a few years later, and the specimen ultimately finished up in the British Museum (now called the Natural History Museum), where it can still be seen today.

The first study of the specimen was made by Sir Everard Home, an anatomist with the Royal College of Surgeons. He published his preliminary findings in 1814. He was initially impressed by its resemblance to a crocodile—the skull had a long tapering snout and lower jaws, crammed full of teeth. He therefore concluded that it was a new species of crocodile. However, when he examined the vertebrae they appeared to be like those of fishes, so he changed his mind and decided it was some kind of fish that formed a link with crocodiles. Home wrestled with the problem for the next five years, eventually deciding the specimen was intermediate between lizards and salamanders. As it happens, he was wrong on all counts. Additional specimens were subsequently found, many by Mary Anning, and it was left to others to recognize that they represented a new group of reptiles. They were named ichthyosaurs, meaning "fish-lizard," because of their superficial resemblance to fishes. Like fishes, ichthyosaurs have forefins and hindfins instead of arms and legs, numerous ribs, and a streamlined body. A curious feature, seen in most of the skeletons, was for the vertebral column to be kinked downward in the tail region. Sir Richard Owen, one of the greatest anatomists of all times, decided that this was a postmortem effect—the tail was really straight in life. Owen was the paleontological voice of authority, so contemporary drawings depicted ichthyosaurs with straight tails. The life-sized ichthyosaur models constructed under Owen's supervision on the grounds of the Crystal Palace,[10] in Sydenham, were given straight tails too. They were also shown to haul up on land like seals. I have seen at least one ichthyosaur skeleton in which the bend in the tail appears to have been straightened out during its preparation, no doubt under Owen's influence. As we will see, Owen was mistaken about the tail, but errors were rare in his long and outstanding scientific career.

Although Mary Anning's discovery was not the first ichthyosaur that had ever been found, it was the first substantial specimen to be studied and reported to the scientific community.[11] Dozens of other ichthyosaurs have since been collected from Lyme Regis and its environs, but this world-renowned ichthyosaur site was soon to become eclipsed by the Holzmaden area of Germany. The quarries in this area of southern Germany have been worked for their building stone for over two centuries and have yielded some of the most remarkably well preserved ichthyosaurs that have ever been found. Shortly after Owen's death, skeletons began to be found in which the body outline had been preserved as a thin carbonaceous film. These specimens, which displayed their beautifully streamlined shape, revealed that ichthyosaurs had a dorsal fin, like that of dolphins and sharks. It also showed that the bend in the tail was natural, and that the down-turned vertebral column formed the skeletal support for the lower lobe of a deep crescentic tail.

Some of the Holzmaden specimens have the remains of developing embryos inside the body cavity. In one particular specimen a juvenile has been

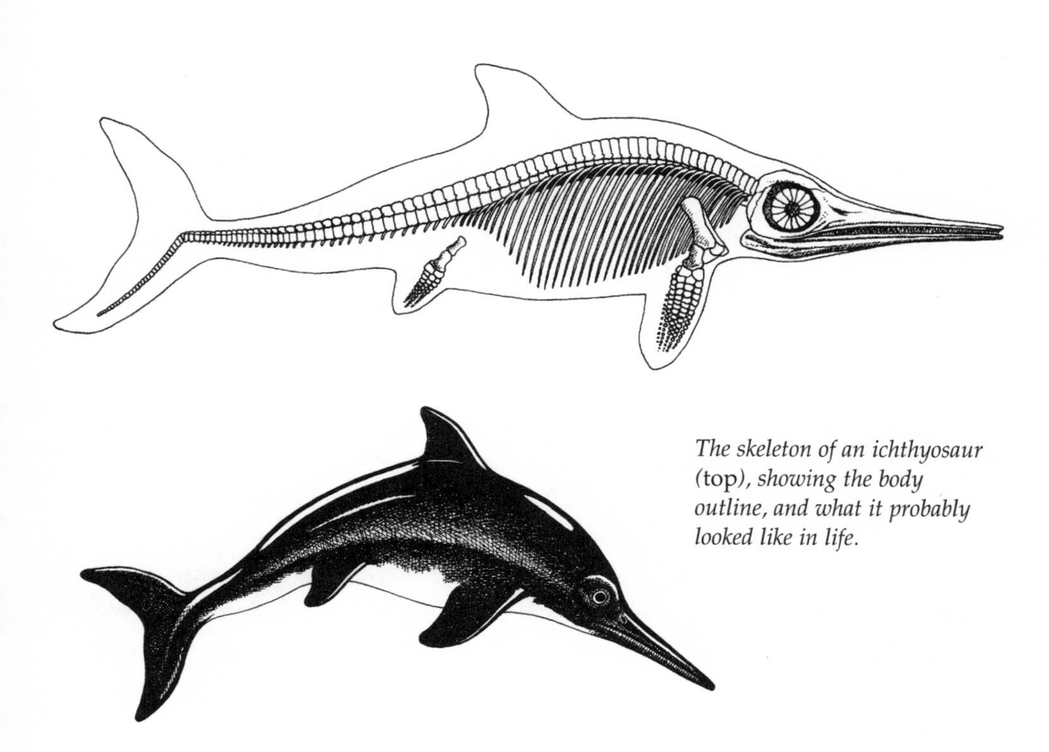

*The skeleton of an ichthyosaur (*top*), showing the body outline, and what it probably looked like in life.*

preserved at the very instant of birth, with most of its body free and only its head still engaged in the birth canal.[12] Like cetaceans, ichthyosaurs appear to have been born tail first, probably to prevent premature drowning during birth. The occurrence of skeletons containing preserved embryos is very rare in the fossil record and permits the sex of the adult to be established unequivocally. It also allows comparisons to be made between mature and immature individuals that unquestionably belong to the same biological species. This again is a rare occurrence.

What sort of animals were ichthyosaurs? Most of them had numerous teeth, usually sharply pointed, which intermeshed when the jaws closed, top with bottom ones, leaving no gaps. Such a dentition, found today in dolphins and seals, is indicative of a diet of fishes and other slippery animals, like squid. This has been confirmed by the fact that several specimens were found with the remains of fishes and squidlike belemnites in the stomach region. We have seen that dolphins, and other toothed cetaceans, find their prey by echolocation. Might ichthyosaurs have been able to do the same? Underwater echolocation requires that the bony capsules that house the two inner-ear structures be acoustically insulated from each other. Cetaceans achieve this with a layer of spongy material that prevents bone-to-bone contact. Our ear capsules are not similarly insulated, which helps explain why we have such difficulty in trying to determine sound directions underwater. The buzz of an outboard motor, for example, seems to come from all around when our head is beneath the surface. Ichthyosaurs also have bone-to-bone contact between their capsules, so echolocation can be ruled out. One of the most prominent features of an ichthyosaur skull is the large eye socket, and these animals probably located their prey by sight. Judged from our scant knowledge of their brain, olfaction was probably of less importance because that part of the brain associated with the sense of smell was not particularly well developed.

Ichthyosaurs were active predators. Being reptiles, they were air-breathers, and, like dolphins and whales, they would have to return to the surface to breathe after diving. They bore their young alive, like cetaceans, and would have no reason to haul ashore like seals and turtles. Nor were they likely to have beached themselves, as do killer whales, to seize prey on the shore.

There were many different kinds of ichthyosaurs, as there are of cetaceans. One of the smallest ichthyosaurs, *Mixosaurus,* which is Triassic in age, was about the size of a salmon, but most genera were much larger, averaging the size of dolphins and porpoises (4–7 feet; 1.3–2.2 m). Among

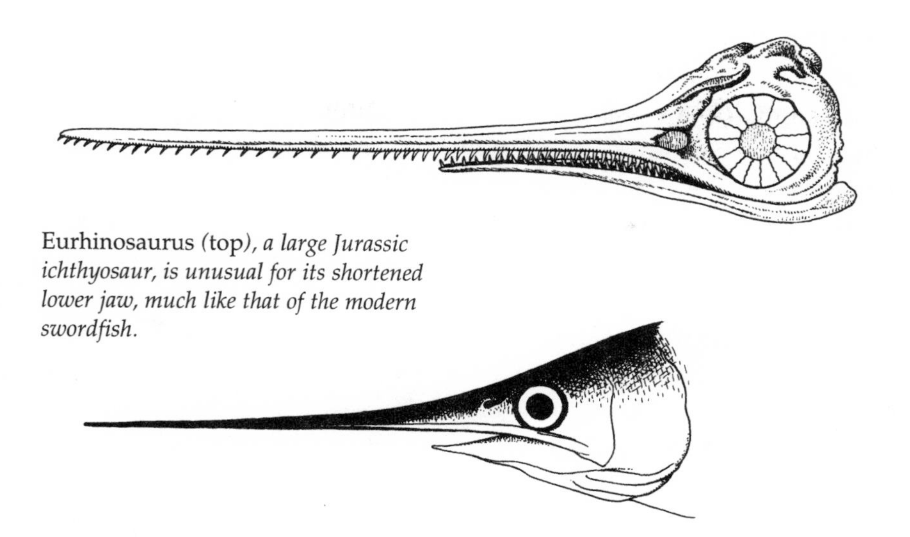

Eurhinosaurus (top)*, a large Jurassic ichthyosaur, is unusual for its shortened lower jaw, much like that of the modern swordfish.*

the most common of these were *Ichthyosaurus* and *Stenopterygius,* both from the Jurassic. Some were larger, like *Temnodontosaurus,* also from the Jurassic, which could reach the size of a killer whale (30 feet; 9 m). *Eurhinosaurus,* from the Jurassic of Europe, was one of the most specialized. This large ichthyosaur, as big as a killer whale, is remarkable for having a short lower jaw and a greatly extended rostrum, much like the modern swordfish. But whereas the swordfish is entirely toothless, the snout and lower jaw of *Eurhinosaurus* are both armed with teeth. Furthermore, the snout of the swordfish is flattened, from above to below, so that its edges are razor sharp, but there is no flattening in *Eurhinosaurus.* Presumably *Eurhinosaurus* was a fish-eater like the swordfish, using its rostrum as a weapon to incapacitate its prey.

The biggest ichthyosaur was *Shonisaurus,* a Late Triassic giant with an estimated length of as much as 49 feet (15 m). This is as large as a sperm whale. *Shonisaurus,* known from a number of incomplete skeletons found in Nevada during the 1950s and early 1960s, has little in common with later ichthyosaurs. Aside from its gigantic size, it is unusual for having a long slender skull with the teeth apparently confined to the front portion. The teeth also appear to be very small compared with the skull: the largest skulls are about 9 feet (2.8 m) long, but the teeth are probably not more than about 2 inches (5 cm) high. Much of the length of the tooth was set in the bone, so there would have been only a short crown showing above the gum line. Ichthyosaurs typically have their teeth set in an open groove, but in *Shonisaurus* they are set in individual sockets.

What would constitute the food of an ichthyosaur with small teeth that were confined to the front of the jaws? Whales that feed on squid have this type of dentition; pilot whales, for example, have just a few teeth, confined to the front of the mouth. The sperm whale has teeth only in its lower jaw, and although the teeth are large, most of their length is embedded in the jawbone. Furthermore, they do not get their teeth until they are older than ten, but they still manage to catch squid before reaching that age, so teeth cannot be so very important to them. It therefore seems likely that *Shonisaurus,* with its poorly developed dentition, fed on squid and similar animals without backbones. But what about the next-largest ichthyosaurs, the killer whale–sized *Temnodontosaurus*?

The teeth of *Temnodontosaurus* were about the same size as those of *Shonisaurus,* but since the skull was smaller—about 6 feet (1.8 m) compared with 9 feet (2.8 m)—they appear somewhat larger. Some of the teeth have cutting edges, and some have blunted points, both features suggesting a diet that included more substantial items than fishes. There are three species of *Temnodontosaurus.* Mary Anning's first ichthyosaur belonged to the species *Temnodontosaurus platyodon,* distinguished by a fairly long snout. While *T. platyodon* may have preyed on other marine reptiles, it seems likely that the second English species, *T. eurycephalus,* occupied such a niche. As is so often the case in paleontology, the species is known from a single specimen. It is also far from complete, being an isolated skull, with no trace of the rest of the animal, so little is known about the species. The skull is large—just over 3 feet (1 m) long—and the short snout and lower jaw are both remarkably deep, indicative of a powerful bite, somewhat reminiscent of the large predatory dinosaurs. The teeth are robust, with short crowns that have cutting edges, and long roots to anchor them firmly in the jaws. The teeth are surprisingly sharp, with no obvious wear facets, suggesting they were used for slicing and cutting through flesh and not for crunching through bones. Perhaps they fed on large marine reptiles the same way that killer whales feed on large whales, by stripping off large slabs of skin and flesh.

Killer whales reach a maximum length of about 32 feet (9.7 m), which is about the upper size limit for *Temnodontosaurus.* Until recently it was thought that *Shonisaurus* was the only ichthyosaur that greatly exceeded this size range, but we now know that a second ichthyosaur also reached sperm-whale proportions. However, unlike *Shonisaurus,* this giant may have hunted other ichthyosaurs. So far we know very little about it, and the trail leading to its discovery, which began in the United States and ended in England, involved some detective work in dusty museum collections. I found the first clue, quite by chance, while visiting the Academy of Natural Sciences, in Philadelphia, to give a talk on ichthyosaurs. I thought I would maximize my time during my brief visit by having a quick look through its ichthyosaur collection. I had already seen the collection before, but this time my attention was focused on large specimens, in anticipation of an upcoming visit to England and Germany for the same purpose. As I was working my way through the specimen drawers, I came upon a large discoidal bone—almost 1 foot (272 mm) across. The specimen, from the Jurassic of Lyme Regis, had been identified as a coracoid bone, which seemed reasonable. The *coracoid* is part of the shoulder girdle, and the general shape, and especially the size, of this particular bone were consistent with its being part of the shoulder of a large ichthyosaur like *Temnodontosaurus.* However, when I looked more closely, I began to suspect that it might not be a coracoid at all, but a quadrate bone instead. The *quadrate* is one of a pair of bones at the back of the skull to which the lower jaws are hinged. If this really was a quadrate, the ichthyosaur in question must have been enormous because this was one very large skull bone. I borrowed the specimen and took it back to Toronto to compare it with some quadrate bones there. Sure enough, it turned out to be a quadrate. To give some idea of how big the skull must have been: There is a 6-foot-long (1.8 m) skull of *T. platyodon* in the Natural History Museum, London, whose quadrate is 8 inches (200 mm) long, compared with the Philadelphia specimen's 11 inches (272 mm). This is 36 percent bigger, so if the isolated quadrate did belong to *T. platyodon,* its skull would be just over 8 feet (2.5 m) long, assuming similar proportions. The entire animal would be at least 36 feet (11 m) long, which is larger than any other specimen of *Temnodontosaurus.* This unexpected discovery alerted me to the possibility that there might be other, and possibly even bigger, ichthyosaurs lurking in museum drawers.

London's Natural History Museum has a very large ichthyosaur collection, mostly from England, and predominantly from Lyme Regis. I got started on my search for giant remains and came across a massive shoulder blade, almost 18 inches (444 mm) long. This was considerably larger than any other I had ever seen from England or from Germany, and I estimated that it belonged to an ichthyosaur that was about 46 feet (14 m) long. Then I found a dinner plate–sized vertebra, over 8 inches (205 mm) in diameter, which would have scaled up to about a 50-foot-long (15 m) monster. But the biggest surprise was still to come.

Temnodontosaurus does not have very large teeth, and the largest ones I have measured are just under 2.5 inches (58 mm) long. Furthermore, the teeth do not appear to relate to the size of the individual, small skeletons having teeth about the same size as those of large ones. As I continued pulling open drawers, I started finding some large teeth, and they seemed to get bigger with every new drawer I opened! The largest of them all, a pair of intermeshing teeth set in a block of stone, had a length of more than 4.5 inches (116 mm), twice the size of the biggest *Temnodontosaurus* teeth. The crown, which has cutting edges, occupies less than one-third of the entire height of the tooth, whereas it accounts for over one-third of the height in *Temnodontosaurus.* The teeth have most in common with those of *T. eurycephalus,* but they are twice as large. The teeth obviously belong to some entirely different kind of ichthyosaur—probably to the same type that the giant quadrate, shoulder blade, and vertebra belonged. One day, perhaps after a heavy winter storm has brought down more of the crumbling sea cliffs along the Dorset coast, one of the local collectors will find more of this enigmatic giant.

Although the largest ichthyosaurs were as big as many of the mysticete whales, there is no evidence that any of them evolved a filtering mechanism to allow them to exploit the vast resources of the plankton. The same is also true for all the other marine reptiles. They include the plesiosaurs, which were contemporaneous with the ichthyosaurs, and the mosasaurs, which evolved late in the Cretaceous, after the ichthyosaur had become extinct. We will see in the next chapter what these marine reptiles were missing.

8

Warfare in Miniature

The night plankton trip is one of the highlights of the marine biology course I teach. The cold waters off Canada's east coast are not noted for the dazzling displays of bioluminescence that are seen in the tropics, but the show is never disappointing. We set out after sundown, a dozen contemplative students in a small trawler, not knowing what to expect. If it is a dark night, with little moonlight, the magic can begin soon after the boat leaves the town lights behind.

"See down there?" A finger points down to the side of the boat. "Look, there's another one. Did you see it?"

Skeptical eyes strain into the dark, unconvinced that there is anything to see besides the sea.

"And another. You must *have seen that one!"*

Somebody admits that he thinks he saw something, but he's not sure.

"They're only small. Tiny pinpricks of blue light. They only fire off for a split second. It's not . . ."

Before the sentence is finished, an electric blue spark flashes and tumbles along the ship's side. Then it is gone, but not before everyone has seen it.

Now that everyone knows what to look for, the students are finding light shows of their own. Some of them hang over the back of the boat and see the sapphire explosions dancing in the screw wash. It is time to collect some plankton. The deck lights are turned on again, and the plankton net is made ready.

The conical net is made of a fine nylon mesh, tapering at the narrow end to a detachable collecting jar. It is lowered over the side and the line paid out. The white net snakes along the surface for several seconds before its 2-foot (0.6 m) mouth fills with water and it sinks into the darkness. The towing speed is kept low to allow the water to filter through the mesh, and to minimize damage to the fragile plankton. The depth of the net is kept at about 10 feet (3 m). After about fifteen minutes the net is hauled back on board. The jar, now full of plankton, is unscrewed and held up to the deck lights. The water teems with animate particles; some are half the size of a grain of rice, but many more are considerably smaller.

Being small is not a prerequisite of being a *plankter*—an individual member of the plankton. Jellyfishes, for example, are planktonic and are commonly the size of saucers, while giants with a diameter of 1 yard (1 m) are found in polar seas. The word plankton comes from the Greek word for wanderer and is used for the myriad of organisms that drift with the currents in the upper layers of oceans and lakes. Most, but not all, plankters are mobile, yet their swimming powers are not sufficient to enable them to swim independently of the currents. So they drift along together, but can swim relative to one another in the plankton.[1] The animals in the plankton constitute the *zooplankton*, while the plants, which are generally much smaller, form the *phytoplankton*. The individual members of the zooplankton are called *zooplankters*, while those of the phytoplankton are called *phytoplankters*.

There is good reason for collecting plankton at night rather than during the day. Aside from being able to see bioluminescence, there are more zooplankters at the surface at night. This is because many undergo a daily vertical migration, dispersing into deeper waters during the daytime and moving up into the surface layers at night. The purpose of this migration seems to be to avoid predators. Many fishes feed on zooplankton, relying on sight to catch their prey. By avoiding the lighter surface layers of the sea during the daytime, zooplankters can reduce their chances of being eaten. The extent to which zooplankters migrate vertically appears to be determined by the abundance of their predators. When herring stocks declined in the North Sea during the 1970s, the zooplankters upon which they feed showed less inclination to migrate than when herrings were plentiful. Although zooplankters are less abundant at the surface during the daytime, you can still catch vast numbers of them using a plankton net.

The deck lights are turned off again and the jar of plankton is taken into the chart-room. It is pitch-black inside, and it takes our eyes several moments to ac-

commodate to the darkness. The person holding the sample can see the bright blue flashes of light that occasionally fire off inside the jar, but most of the others cannot see them because they do not know where to look. At my suggestion the custodian of the jar gives it a good shake, and the entire contents are transformed into a glowing specter. There is a collective gasp of approval. The phosphorescence lasts for several seconds, then all is dark again. The jar is shaken once more, and the phantom light reappears, as if a switch had been thrown. Everyone else wants to try, and the jar is passed from hand to hand. At length the jar of plankton is retrieved and put into a cooler of crushed ice—if it is not kept cold, the minute organisms will soon die.

Back on shore the plankton sample can be examined under a binocular microscope. A few drops are transferred from the jar to a petri dish, using an eyedropper. The lowest magnification is selected, and the petri dish is set down on a black background, with the light shining down into it. The advantage of using black field illumination is that the colors of the plankton show up more vividly. And what an astonishingly beautiful world it is.

The most conspicuous members of the plankton are the copepods. These tiny crustaceans, distantly related to shrimps and crabs, are only the size of a flea (0.5–1.5 mm) but are giants compared to most of the other plankters. They have torpedo-shaped bodies and a pair of long antennae, the first antennae, which they hold out in front of them as they move through the water. There are numerous genera and species, found in fresh-

Copepods, small crustaceans, are the most conspicuous members of the plankton. The largest one shown here is Calanus, *one of the most common.*

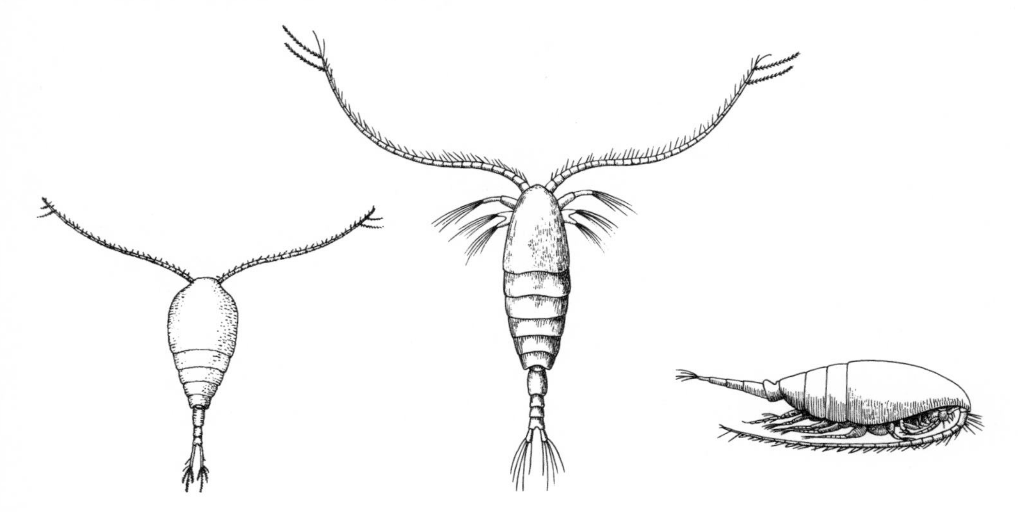

water as well as in the sea. It is easier and more convenient to collect samples from ponds and lakes than from the ocean. Freshwater plankters are also generally easier to maintain in the laboratory than marine ones (unless the lab is beside the ocean), so much of what we know about copepods, and many other plankters, has been learned from studying freshwater ones. However, the freshwater species are apt to behave much like their close relatives in the sea, so the results of freshwater studies are probably applicable to the marine situation.

Copepods have two main ways of propelling themselves through the water. They can cruise along fairly slowly, at speeds of a few millimeters a second or less, or they can lunge, for very short distances, at speeds in excess of 200 or 300 millimeters per second. These remarkably high speeds—the equivalent of a porpoise accelerating to over 800 mph (1,380 km/h)—result from a scaling effect due to their small size. We saw a similar situation on land for insects and other small animals. A second factor comes into play for small things that move in fluids, and this all has to do with viscosity. The effects, which can be quite bizarre, dominate life in the plankton, so we need to consider them briefly before going any further.

Viscosity can be thought of as a measure of how readily one layer of a fluid slips past an adjacent layer. In general, the lower the viscosity, the easier the flow. Pancake syrup, for example, is far more viscous than water and therefore pours more slowly from a jug. If you filled a pool with cold syrup, you would find swimming very laborious and make little headway. This is because the layers of syrup would not slip past one another as readily as they would if they were water, so there would be far more drag (resistance) on your body. Air is a fluid too, as are all gases. We do not normally think of air as being very viscous, but for very small objects, it is the equivalent of syrup. When a shaft of sunlight enters a room, for example, the minute dust particles that dance and tumble in the beam appear to be floating in a viscous fluid. The same is also true for the small plankters moving about in water. Their situation is comparable to what it would be like for us swimming in syrup. One of the unusual consequences of this high-viscosity situation is that when a copepod stops swimming it comes to a complete halt, as if it had hit a brick wall. Fishes, in contrast, coast along for a little way after they have stopped swimming.

Copepods, like lobsters and other crustaceans, have several pairs of small feeding appendages close to their mouths. Included among these is a second, and very short, pair of antennae. By beating these small antennae, and their other feeding appendages, copepods draw a current of water to-

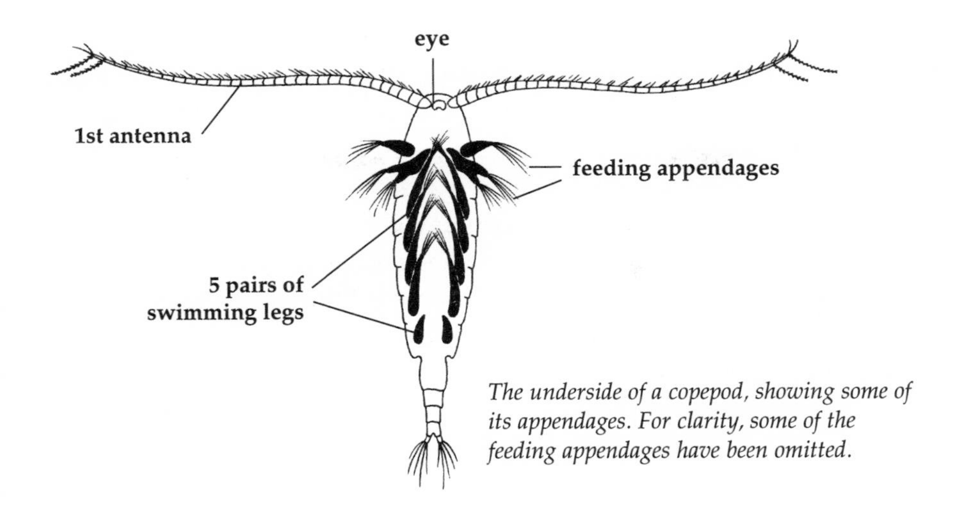

The underside of a copepod, showing some of its appendages. For clarity, some of the feeding appendages have been omitted.

ward their mouths. Small particles—mostly other plankters—become caught up in the water current, and those suitable for food are captured by the feeding appendages, which are edged with numerous bristles. The circular currents set up in the water on either side of the body by the feeding appendages also propel the copepod slowly forward.

In addition to the feeding appendages, copepods have five swimming legs. These are normally kept tucked up beneath the body, pointing forward. But when they are kicked backward (together with the large, first antennae), the animal lunges forward. These lunging bursts of speed, which are used to escape from predators, are of brief duration. Predatory copepods use their short bursts of speed for capturing prey.

Some copepods are herbivorous and cruise along through the water gathering up phytoplankters. The smallest food particles, those about the size of a human red blood cell, are captured passively, that is, they are collected on the feeding appendages and then moved to the mouth. Larger particles are actively seized by individual feeding appendages and transferred to the mouth. Several studies have shown that other copepods prefer to capture motile (capable of moving) plankters rather than nonmotile ones, and it seems that the copepod detects the former because of the disturbances they generate in the water by their swimming movements.

When you look at plankton under the microscope, it is easy to forget that the samples have been concentrated by the collecting process. The individual plankters are far more widely dispersed than this in the sea, so copepods have to process large quantities of water to obtain sufficient food. Their feeding appendages are beating away all the time to keep the stream of water flowing toward the mouth. But every so often they may have to stop, to manipulate a wayward particle, or to deal with one that is particularly large or awkward. As soon as the appendages stop beating, the flow of water comes to an abrupt stop, because of the high-viscosity situation. One advantage of this situation is that any particles that were close to entering the mouth before the flow ceased will still be in the same position, so nothing will be lost. But the disadvantage is that the copepod has to do extra work to start the water flowing again. And the smaller the copepod, the more work it has to do, relative to its size.

Most copepods are omnivorous, feeding on plants as well as on animals, while others are predatory. One of the most common copepods, *Calanus,* which is also one of the larger ones, is an omnivore. When feeding on phytoplankton it acts passively, gathering them all up, but when it detects suitable prey in the water it becomes an active predator, scooping and grasping at them with its mouthparts, and lunging toward them, if necessary, to close the gap. Since copepods feed as effectively in darkness as during daylight, they are obviously not using their eyes to locate their prey. So how are they doing it? Some clue is given by the sort of prey that attract their attention, but before discussing this I need to say something about larvae.

While many of the animals in the plankton are adult, a large portion are immature. Indeed, the plankton is a nursery ground for the myriad of species that pass through a larval phase before developing into adults. These include most of the animals of the seashore, like barnacles, sea urchins, bristle-worms, anemones, starfishes, corals, molluscs, crabs, sea squirts, and sponges, as well as numerous fishes. Larvae are usually strikingly different from their parents, a familiar example being the caterpillar, which could hardly look less like a moth or a butterfly. One of the reasons for passing through a larval phase is to gain weight and grow, and caterpillar's, with their voracious appetites, are good examples of this. Another vital biological role that larvae fulfill is the dispersal of the species, and this is especially important for sedentary animals like barnacles and sponges. Although larvae look so different from their parents, the larvae of the same groups of animals look very much alike, especially in their early stages. Crustaceans, for example, pass through an early larval phase called a *nauplius*—which looks a bit like an egg with legs—and the nauplius larva of a crab or a barnacle looks very much like that of a shrimp or a copepod, even though their respective parents look so different.

Let's now return to *Calanus* and the prey that most attracts its attention.

Rather surprisingly, *Calanus* has a predilection for the larvae of copepods, and since it is itself one of the more common genera, it must consume many of its own kind. The larger larvae are the easiest for the copepods to locate, probably because they make the biggest disturbances in the water. But how are these vibrations detected? The large antennae, which are covered with bristles, are stretched out in front of the copepod as it moves through the water, and are the most likely detectors of water-borne vibrations. This suggestion is supported by experiments in which copepods have had their antennae amputated. Their ability to gather phytoplankton is unimpaired, because they can still generate feeding currents, but they are far less able to capture prey. There is, however, an alternative explanation that fits the facts. Jeannette Yen, a plankton specialist at the State University of New York, suspects that the inability of the amputees to capture prey has nothing to do with the loss of sensory function. Instead, she thinks it is simply because copepods use their large antennae to seize their prey.

When I was a student, more years ago now than I care to recall, my classmates and I usually got to see only dead plankton, preserved in formalin. Formalin-fixed copepods look drab and gray, but the living animals are dazzlingly attractive. Viewed with reflected light against a black background, their thin exoskeletons appear opalescent and glow with all the colors of the rainbow. But if the light is changed so that it now shines through them, their internal organs can be seen through their transparent bodies. The most prominent organ is the gut, which often contains small emeralds—the freshly ingested phytoplankton of the animal's last meal. But for all their beauty, live copepods can be frustratingly difficult to observe. Aside from the incessant beating of their mouthparts, which causes them to move all the time, they occasionally dart completely out of the field of view, by the spontaneous kicking of their swimming legs. They can be confined by putting them on a microscope slide and dropping a coverslip (a thin wafer of glass) on top of them, but then their actions are no longer normal. These problems have frustrated researchers for over a century, but with the advent of high-speed movie cameras, as well as video cameras and computers to analyze the images, it became possible to follow their unrestricted movements. In one study, which plotted the fates of potential prey items in their path, it was revealed that copepods have an attack zone around them. Prey that fall within this zone, which has a radius of between two and three body lengths, can be detected and are liable to be attacked. But the probability of attack is not the same all around the circle. The biggest and juiciest of targets are perfectly safe so long as they stay directly behind a hunting copepod. But those in the front, especially if they lie to the left or the right, are the most vulnerable to attack. Current research by Yen and her associates shows that when potential prey are drawn into the circular currents in these two areas, the water movements set off their escape response. The potential prey's escape response, in turn, sets up small disturbances in the water which alert the copepod to its presence.

Copepods and their prospective prey play a cat-and-mouse game, like a pair of enemy submarines. They are blind, and the attacker detects its quarry by picking up its disturbances in the water. But there are fundamental differences between the water disturbances generated by submarines and those produced by plankters, due to the difference in scale. When a submarine, or a fish, moves through the water, it is large enough to generate pressure waves, and these spread out, over long distances, at the speed of sound. These *far-field sounds,* as they are called, lose energy in direct proportion to the distance from the source. The signal strength of a large moving target at a distance of 100 feet (31 m), for example, would be half of what it would be at 50 feet. Plankters like copepods are far too small to generate pressure waves, but they do cause local swirling patterns in the water in their immediate vicinity. These local flows are called *near-field displacements.* Large objects like submarines and fishes churn up the water as they go, and the turbulence they generate prevents any near-field displacements from being formed. Small plankters, in contrast, slip smoothly through the water, and the laminar (smooth) flow of the water around them allows near-field displacements to form. You have probably seen near-field displacements without realizing it. When the sun shines down on a lazy stream, you can often see not only the swirls that form around a projecting plant stem, but also a reflection of these swirls on the bottom. But if the stream is fast moving, the turbulence of the flow does not permit these swirling patterns to form. Copepods can detect the near-field displacements of their prey, but these signatures in the water do not travel very far before they die away. This is because their signal strengths decrease with the cube of the distance. Therefore a local flow in the water two millimeters away from a copepod larva would be only one-eighth as strong as it would be at a distance of one millimeter (the distance is doubled, and $2 \times 2 \times 2 = 8$). This explains why copepods can detect their prey only within a range of two or three body lengths, which is just a few millimeters. Copepods can detect large prey more easily than small ones because bigger targets produce larger near-field displacements, so the copepod's antennae are more likely to intercept them. There is, however, no difference in range over which large and small objects can be detected, because both are determined by the cube of the distance traveled by the water disturbance, not by their size.

What strategies are most adaptive for predatory copepods, and their prey, to enhance their chances for success? The hunter could improve its chances of detecting prey by having long antennae, to cover a broader front, and these should be well supplied with sensory bristles. Some copepods have much longer and more bristly antennae than others, and such features might reflect their predatory inclinations. In *Calanus,* for example, the antennae are longer than the body and have numerous bristles, with extra long ones at the end. *Microcalanus,* in contrast, which is a herbivore, has antennae that are shorter than the body and these have few bristles. However, there are exceptions, like *Eucalanus pileatus,* which has long feathery antennae but which is primarily a herbivore.

The prey species can reduce their chances of being caught by eliminating or reducing the size of the disturbances they make in the water. The freshwater copepod *Diaptomus,* which grazes on phytoplankton, modifies its behavior to achieve these ends. It spends most of its time alternating periods of slow swimming (up to 0.4 mm per second) or rest with periods of rapid swimming. During these intermittent lunges it reaches speeds of up to about 100 millimeters per second, the objective possibly being to reach fresh patches of phytoplankton. The large disturbances these lunges create in the water attract the attention of any predatory copepods that are in the vicinity. However, when *Diaptomus* detects the presence of a predator, like the copepod *Limnocalanus,* the periods of rest are extended. In one particular instance a *Limnocalanus* cruised within about two body lengths of a *Diaptomus* without noticing it, and it was only when the herbivore began lunging away that the predator responded and gave chase.

Many copepods, including *Acartia,* move in short rapid bursts, generally in an upward direction, followed by a passive sinking phase, under the influence of gravity. Their motion is therefore predominantly in a vertical direction. *Euchaeta* is a predatory copepod that includes *Acartia* among its prey. It swims through the water, searching for its prey with antennae that are well equipped with long sensory bristles. Since the local disturbances left in the water by the prey can only be detected over distances of a few millimeters, the predator has to pass very close to the prey to stand any chance of interception. How can the odds be improved? Theoretically, the chances of intercepting a target that is moving predominantly in one direction are increased if the attacker moves at right angles to that direction. A German U-boat, for example, probably had a greater chance of intercepting a ship if it sailed at right angles to its estimated course, rather than approaching on a parallel course. The hunting strategy of *Euchaeta* is therefore to swim steadily on an essentially horizontal course.

Details of this strategy have been revealed using a laser-illuminated video-monitoring system, developed by Rudi Strickler of the University of Milwaukee. The species studied, *Euchaeta rimana,* is a fairly large copepod (2.5 mm) that preys on the much smaller *Acartia fossae* (0.7 mm). The predator swims fairly steadily, at an average speed of 7 millimeters per second. This may sound painfully slow, but it is a reasonable speed relative to the size of the animal and is about three body lengths per second. This is the equivalent of a 6-foot-long (1.8 m) porpoise swimming at about 4 mph (6.4 km/h)—a good cruising speed for a porpoise. The cruising *Euchaeta* propels itself using its short antennae, keeping the long antennae well forward throughout, so that the sensory bristles are kept clear of its own wash. Cruising is only interrupted to make turns, and these are completed in less than sixty milliseconds. Turning has to be done quickly because it makes swirls in the water that flood over the antennae, temporarily "blinding" the predator.

When a target is detected, *Euchaeta* lunges, using its swimming legs and long antennae, achieving speeds of up to 142 millimeters per second, equivalent to a porpoise sprinting at 230 mph (370 km/h). It was not possible to give an average swimming speed for its prey, *Acartia,* because of its intermittent swimming, but it reaches speeds of up to 87 millimeters per second during its short lunges. This is only about 60 percent of the top speed of *Euchaeta,* suggesting that the predator has the advantage of speed. However, as we saw in chapter 1, acceleration is of more importance than top speed in predator-prey encounters, at least in the terrestrial environment. There are no comparative data for the respective accelerations of these two copepods, but *Acartia* would have an advantage if it detected the presence of *Euchaeta* and leaped away before the predator launched its attack. In some encounters it seems that the prey is indeed the first to detect the other, lunging off before the predator pounces.

Another species preyed upon by *Euchaeta* is the small copepod *Corycaeus.* Like *Acartia,* this plankter makes short lunges, usually upward, followed by a passive sinking phase. In one of the sequence recorded, a 2.5-millimeter-long *Euchaeta* lunged toward a 0.6-millimeter-long *Corycaeus,* at a speed of 43.5 millimeters per second. This action caused its potential prey to leap away at 64.5 millimeters per second, the equivalent of a porpoise accelerating to 430 mph (690 km/h). Both copepods continued swimming slowly after this first attack, but the predator then made a second lunge. In response, its prey rocketed away at a staggering 129 millimeters per second, which is equivalent to a porpoise reaching 860 mph (1,380 km/h)! This lunge carried the potential prey a distance of 8 millimeters, or thirteen body lengths away, placing it safely outside the detection range of its attacker.

An unusual feature of *Euchaeta rimana* is that adult males do not feed; their mouthparts are degenerate, and their antennae lack the well-developed bristles of the female. The only objective of their short lives is to find a female and then mate. To this end, they adopt a swimming pattern similar to the prey species of the female, moving in a predominantly vertical direction by interspersing brief upward leaps with passive sinking phases. In contrast to the prey species, though, the swimming segment is at a speed of only 7.5 millimeters per second, comparable to the female's, rather than at a rapid dash. This generates a smaller disturbance in the water, making the male less obvious to potential predators. He counts on being intercepted by the female, and since he has a limited fuel supply to sustain him during the waiting period, anything that can be done to reduce his energy expenditures is at a premium. Compared with the female, the male is more slender, which reduces the drag forces on his body during the upward swim. He also flips over during the sinking phase, so that his body is horizontal, increasing the drag on his body and prolonging his descent. These two features serve to reduce the male's energy expenditures, thereby increasing the time he can remain in the plankton to encounter the female. Although the male has such a modest swimming speed, he is able to lunge away from predatory attacks at speeds of up to 360 millimeters per second, or 150 body lengths per second, which is the equivalent of a porpoise dashing away at 600 mph (965 km/h).

Less numerous than copepods, but still quite abundant in the plankton, especially in freshwater, are the cladocerans. These little crustaceans are generally smaller than copepods and are globular rather than elongate. Their most distinguishing feature is that their bodies are encased in a protective casing. This comprises two valves, or shells, attached to the animal's back, that extend

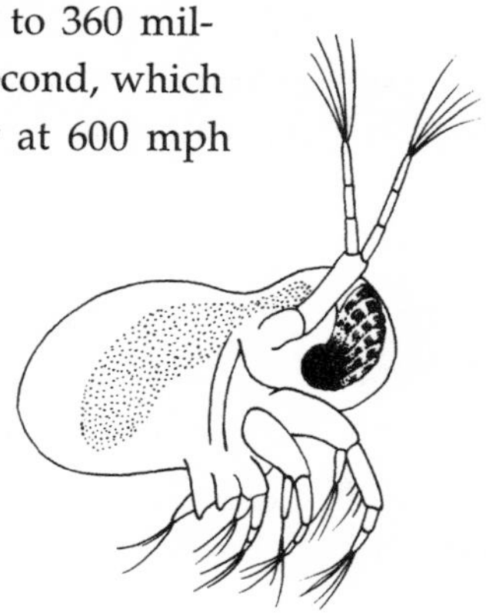

Podon *is a common cladoceran in marine plankton samples.*

down its sides and meet, but do not join, in the middle. The head is free of the bivalved carapace; so too are the antennae, which are the sole swimming organs. The antennae are short, usually branched, have numerous bristles, and can be tucked away in a groove. Cladocerans are preyed upon by copepods (among other predators), and since they are much slower, they cannot outrun them. They must therefore rely on their defensive carapace for protection. A number of copepod attacks on freshwater cladocerans have been recorded,[2] and these revealed similar patterns of offense and defense. The copepod typically detects the cladoceran before the cladoceran is aware of its presence, so the initiative is usually taken by the predator. All attacks involve a lunge by the copepod, usually toward the head region of the prey, using the powerful kicking movements of its swimming legs. The strike is executed with the mouthparts extended, and once contact is made, these appendages are retracted in an attempt to grab hold of some protruding part of the cladoceran's body. The cladoceran responds by retracting its antennae, tightly clamping the two halves of its carapace together, and sinking passively in the water. If the copepod fails to seize its prey, or if it loses contact with it sometime during the attack, the cladoceran has a good chance of slipping quietly away. This is because the cladoceran's compact body creates a minimal water disturbance as it sinks, so its attacker cannot find it. The copepod's strategy is to make repeated grabs at its prey during the attack. If it succeeds in gaining a hold, it wraps itself around the cladoceran's body in a deadly embrace. The attacker then tries prying the two halves of the carapace apart, so that it can insert an appendage inside. Once the carapace has been invaded, the copepod rips out the gut and other soft organs and eats them, discarding the head and the rest of the body. If the cladoceran is too large to be embraced by the copepod, the predator makes repeated attacks on the outside, inflicting as much damage as possible, until it can get inside the carapace. Cladocerans that manage to escape the initial attack have a greater chance of being relocated if they have been injured. Presumably this is because predatory copepods can detect the scent of the tissue fluids leaking out into the water. Herbivorous species certainly have an olfactory sense and use it to swim toward patches of phytoplankton.

When copepods (and other zooplankters) graze on phytoplankton, the phytoplankters release into the water a chemical called dimethylsulphide, which is a product of normal metabolism. The release may be caused by the damage done to the plant cells when they are eaten, but regardless of its origin, large quantities of the chemical are produced in the oceans every day. Certain seabirds feed on plankton, notably some of the members of the tube-nosed, or procellariiform, group (which includes albatrosses and petrels). These birds have an uncanny knack for finding local concentrations of food while patrolling over vast expanses of seemingly featureless ocean. It has been found that some of them are able to detect the smell of the dimethylsulphide, and certain species, like the storm petrels, are highly attracted to it. Their remarkable ability to home in on likely feeding spots may therefore be entirely due to the chemicals released from the doomed phytoplankters.

Plankton samples, no matter from which oceans they have been collected, almost always contain at least a few arrowworms. They are so named for the way they streak, like arrows, toward their prey. By copepod standards, these slender-bodied animals are huge, almost 2 inches long (about 5 cm). Their scientific name of chaetognath refers to the ring of spines that surrounds the mouth. The spines are sensitive to disturbances in the water. Experiments using a vibrating glass bristle have shown that arrowworms can only detect water disturbances up to a distance of about 0.15 inch (3 mm). Increasing the signal strength of the vibrations does not extend the range, which is precisely what is expected for near-field displacements. However, an unexpected finding for one of the species (*Sagitta hispida*) was that it was most sensitive to vibrations in the vicinity of 150 Hz, whereas the frequencies normally associated with zooplankters are in the range of 20–50 Hz. The researchers had no explanation for this, except to suggest that copepods, upon which arrowworms frequently prey, may produce higher-frequency vibrations, but at signal strengths too low for us to detect. Copepods probably generate higher frequencies than 50 Hz when making their high-speed dashes from danger. Arrowworms are most active at night, when they probably do most of their hunting.

If you peer down a microscope at some living plankton, you can be forgiven for thinking you are visiting another planet. The teaming life-forms, as alien as they are beautiful, zoom in and out of the field of view, propelled by unseen forces. And as your eyes scan the bustling copepod empire, a diaphanous creature comes into view, half gliding, half floating, like a dream. The transparent bell, no more than a few millimeters in diameter, pulsates rhythmically, driving the animal forward by jet propulsion. It trails four long tentacles behind it, but some other kinds have numerous tentacles. These plankters belong to a group of animals collectively known as cnidarians or coelenterates, and which includes sea anemones, corals, hydroids, and jellyfishes. Many of the group are sedentary; the hydroids, for example, look like small plants and are commonly found on the seashore. Most of the hydroids exist in two forms: the polyp, which is the sedentary plantlike form, and the free-swimming medusa. The medusae, which bear the gonads, are the transient sexual phase of the life cycle and serve to distribute the species. Jellyfishes belong to a different subgroup from the hydroids, where the medusoid phase is the main part of the life cycle. Many of the medusae seen in plankton samples, especially those collected close to land, are the sexual phases of hydroids, whereas others are jellyfishes.

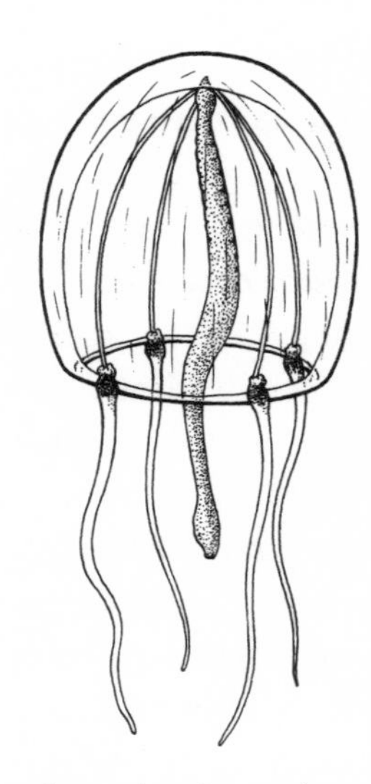

A free-swimming medusa, named Sarsia, *whose bell sometimes reaches a height of about 10 millimeters.*

For all their innocent beauty, medusae are predators, capturing and killing their prey using minute harpoons, called *nematocysts* or stinging cells. These microscopic structures (found in other coelenterates too) are embedded in the outer layer of the medusa, primarily in the tentacles. When they are set off, usually by physical contact, as when an animal brushes up against one of the tentacles, a barbed thread is fired out. The barb punctures the skin or exoskeleton of the prey, inoculating it with venom and securing it at the same time. Prior to discharge, the harpoon is turned inside out and is stored, neatly coiled, inside the body of the cell. A small trigger, called the *cnidocil*, projects from the cell, and when this is tripped it causes an in-flooding of water that makes the thread and harpoon instantly swell, resulting in the discharge. A mixture of substances is found in the venom, including a neurotoxin, histamine (which causes a rash in humans), and enzymes. Small animals are rapidly immobilized, and some coelenterates can harm large animals

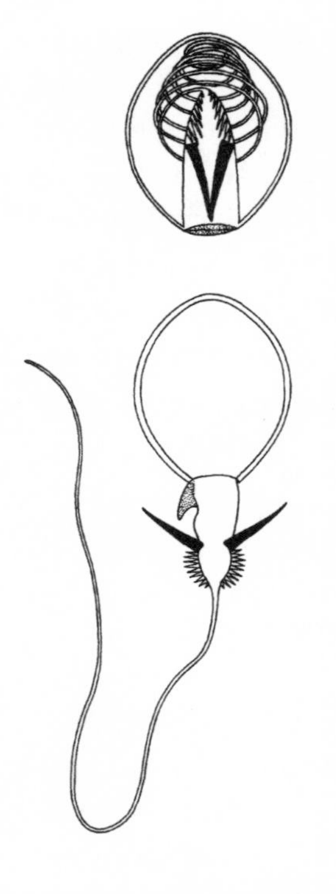

A nematocyst, before (top) *and after being discharged.*

too. Many of them can give a painful sting to humans. At least one, the jellyfish called the sea wasp (*Chironex fleckeri*), which is found in Australian waters, has a sting that is often fatal. The venom of this species affects the heart and is so toxic that death occurs within minutes. Since jellyfish are transparent, they are seldom seen by swimmers. The long tentacles that many of them trail in the water are similarly difficult to see. I am always ultracautious when swimming in tropical waters, but on one occasion in the Pacific I was painfully stung by a jellyfish that I never saw. My first indication of trouble was a sudden flush of excruciating pain in my arm, as if a kettle of scalding water had been poured over it. When I looked down, I saw that a fine blue thread had wrapped itself around my arm. I tried to pull it off with my other hand, but it was like rubber and just snapped back again, with another burst of pain. I got out of the water as fast as I could and managed to remove the offending tentacle on shore, but the pain lasted for the rest of the day and I had a rash for a couple of months afterward.

Although most animals stay well away from coelenterates and their venomous nematocysts, others are unaffected by them. Some, like nudibranch molluscs, actually feed on them. The nudibranchs are small shell-less molluscs, often brightly colored, which typically live along the seashore. Not only do some nudibranchs feed on sea anemones and hydroids, but a few, like *Eolis*, extract the nematocysts and use them for their own defense. The appropriated weapons are transported to a series of processes, called *cerata*, that cover the back of the animal. It is unclear how *Eolis* is able to selectively save the nematocysts during the digestion of the rest of the coelenterate meal, and how it prevents the nematocysts from being discharged until they are needed.

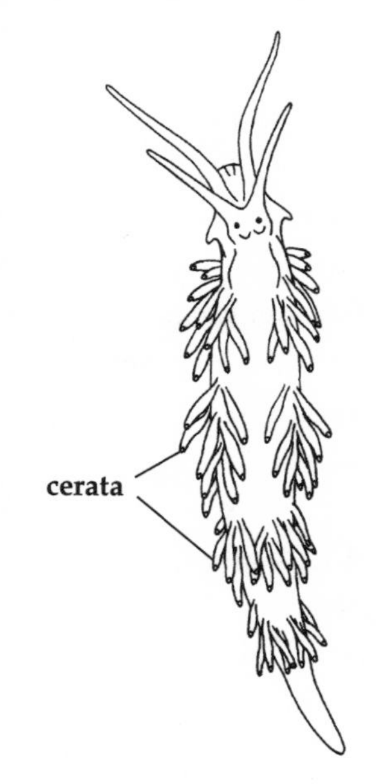

The nudibranch mollusc Eolis, *which feeds on sea anemones, extracts the nematocysts and stores them in the cerata for use in its own defense.*

Next to copepods, the most obvious members of the plankton are the diatoms, which are members of the phytoplankton. They are single-celled organisms, so we have to look at them under higher magnification. This involves putting a drop of the plankton sample on a glass slide and placing a coverslip on top, spreading the water

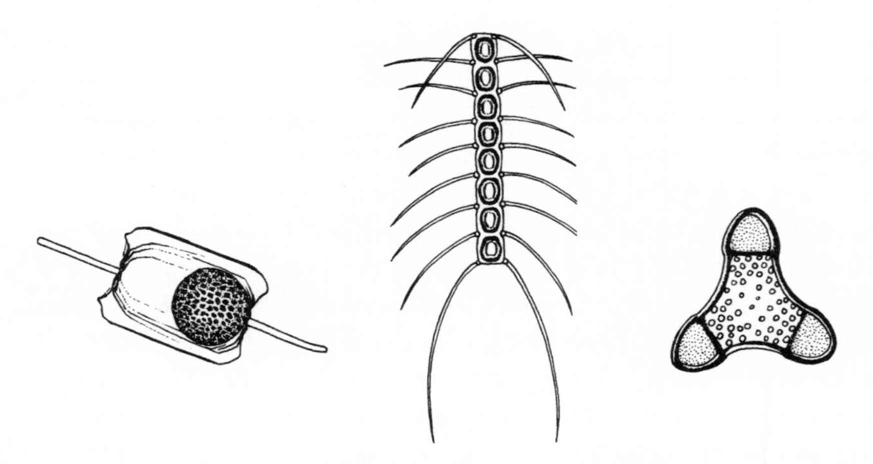

Diatoms have glassy exoskeletons, many with long spines. Some diatoms remain isolated, but others form chains and clumps of individual cells.

out as a thin film. The smallest diatoms are minute, no larger than a human red blood cell, but the largest ones have a diameter of about one-tenth of a millimeter. Diatoms come in a wide array of shapes, and while many species remain as isolated cells, others unite to form chains and clumps. All diatoms have a siliceous (glassy) exoskeleton, and many of them, especially the colonial ones, have long slender spines. These function to increase the drag forces, so the cells, which tend to be denser than the surrounding sea water, sink more slowly. The spines may also discourage herbivores, like copepods, from browsing on them. Like other plants, the diatoms contain chlorophyll and other pigments used in photosynthesis, and these pigments are contained in small bodies called *chloroplasts.* Some chloroplasts are green, but others vary from yellow, to straw colored, to brown.

Whereas most planktonic diatoms do not have motility, most of the other protozoans do. This motility is easily observed under the microscope. And as you admire the diatoms' cut-glass beauty, something zooms into your field of view and streaks out again just as quickly. By reducing the magnification to increase the field of view, you manage to locate the speeding protozoan. It glides effortlessly beneath the coverslip. You try to center on the specimen, increasing the magnification for a closer view, but it keeps moving too fast. If you could slow it down, you would see that it was a ciliate, so called because its surface is covered by small hairlike processes called *cilia. Paramecium,* remembered from high school biology, is a ciliate. Individuals range in size from 20 micrometers to 2 millimeters, so the largest ones can be seen with the unaided eye. Most of them are in the range of 50–500 micrometers. Ciliates, regardless of their size, move at similar speeds—about one millimeter per second. For the smallest ones, this is about fifty body lengths per second, or the equivalent of a porpoise traveling at 200 mph (320 km/h). But for a 2-millimeter-long giant, it is a leisurely pace, like a porpoise cruising along at 2 mph (3.2 km/h).

The cilia have a fairly constant diameter of 0.25 micrometers, or about one-thirtieth of the diameter of a human red blood cell. They are constantly beating, in unison, like the oars of a rowing team, and it is this movement that propels the organism through the water. During the power stroke each cilium is stiff, acting like an oar against the water. When a boat is being rowed, the recovery stroke is made with the oar out of the water, but this is not possible for the cilia because they are immersed in water. If the recovery stroke were made with a stiff cilium, it would cancel out the power stroke, no matter how slowly it moved, because of the high viscosity. The recovery stroke is therefore made with a limp cilium, which is kept close to the body surface to minimize the drag force. Cilia do not always cover the entire surface of the body. Sometimes several cilia are joined, forming a specialized structure. The *cirri*, for example, are stiff bristles, used in locomotion, that are formed from clumps of cilia.

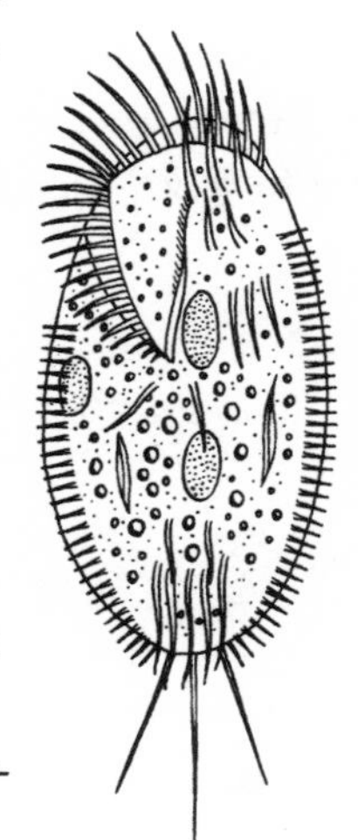

Many ciliates, like *Paramecium*, feed on small organisms, such as bacteria, which they collect passively by generating feeding currents with their cilia. Other ciliates are active predators, hunting prey that may be larger than themselves. These are often other ciliates. Ciliates are constantly on the move. Since none of them have any apparatus for seizing, how do the predatory ones catch their prey? Many ciliates, including the predators, have minute structures called *trichocysts*. These are like miniature nematocysts, except that they fire out a thread rather than a harpoon. Some trichocysts (called toxicists) discharge toxins, and a detailed study of the predatory ciliate *Litonotus* shows how these toxins are used to paralyze

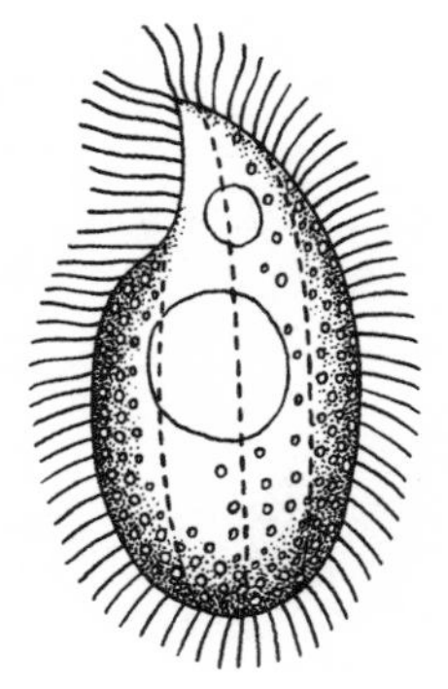

Ciliates are the fastest single-celled organisms in the plankton.

prey. During the study, starving specimens of *Litonotus* were put among some smaller ciliates, called *Euplotes,* upon which they prey. Since ciliates dash around so rapidly, collisions are inevitable. When *Litonotus* collided with *Euplotes,* it immediately fired off numerous toxicists, and the prey stopped moving shortly afterward. The predator's toxins targeted the prey's locomotory cilia, paralyzing them, whereas the other cilia kept beating for about one minute longer. Examination of the paralyzed cilia under the electron microscope revealed that the ends were swollen. Once the prey was completely motionless, the predator relocated it, presumably by chemical cues, and began feeding. *Euplotes* is apparently unable to defend itself against *Litonotus.* However, some freshwater species of the same genus have a novel defense against the predatory ciliate *Lembadion.* When exposed to water containing this predator, the prey grow a pair of winglike extensions from their body, almost doubling their size. This makes them too large to be swallowed. However, the change takes several hours to develop, so it cannot give them instant protection. The transformation is triggered by some unknown substance produced by *Lembadion* that diffuses into the water. Provided the concentration of the substance in the water is maintained, the altered shape persists. It is even passed on to the next generation when the cells divide during their asexual reproduction.

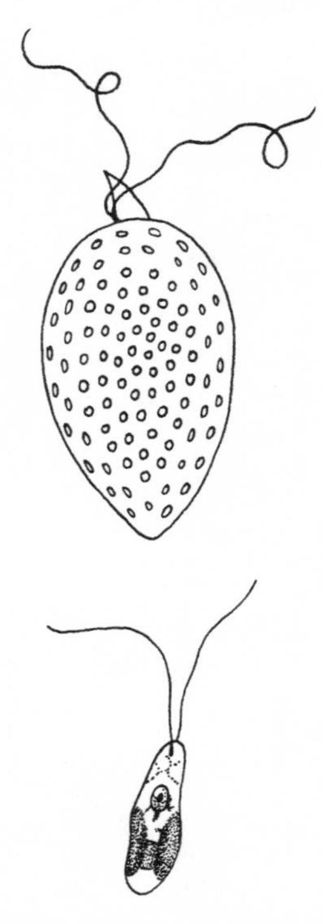

Flagellates move at about one-tenth the speed of ciliates and are about one-tenth as large.

Much easier to watch than ciliates are the flagellates, which travel at about one-tenth of their speed. They are also about one-tenth of their size, ranging from 1 to 50 micrometers. Flagellates are so named for their whiplike motile organ, the *flagellum,* has the same structure as a cilium but is much longer, often longer than the entire body. Flagellates usually have between one and four flagella, and these are undulated to produce thrust, either in one plane or in a helical fashion. While the undulating flagellum may look like the undulating tail of a fish or tadpole, it works on entirely different principles owing to the high viscosity of the flagellate's environment. Because of its small size, you need to use a high magnification to be able to see it properly. As you bring the image into focus, a small flagellate comes into view. The beating of its long flagellum causes its body to wobble slightly, and since the water effectively has the consistency of syrup, the movement is transferred to the surrounding water.

More conspicuous than the highly motile flagellates are the dinoflagellates, most of which have a thick outer casing of cellulose. They have two flagella, one lying in a prominent horizontal groove, the other in a vertical groove which is not always as obvious. Among the most common dinoflagellates is *Ceratium*, a spidery-looking organism that usually has three long slender spikes, though several species have only two. Some species of *Ceratium* are bioluminescent, but the most spectacular emitter of light is *Noctiluca*, a huge gelatinous sphere that reaches about one millimeter in diameter. *Noctiluca* often occurs in dense swarms, especially near coasts, and is the primary cause of bioluminescence in temperate waters. The electric blue flashes seen on the night plankton trip were caused by *Noctiluca*. But why should these organisms invest so much chemical energy in producing these spectacular light shows?

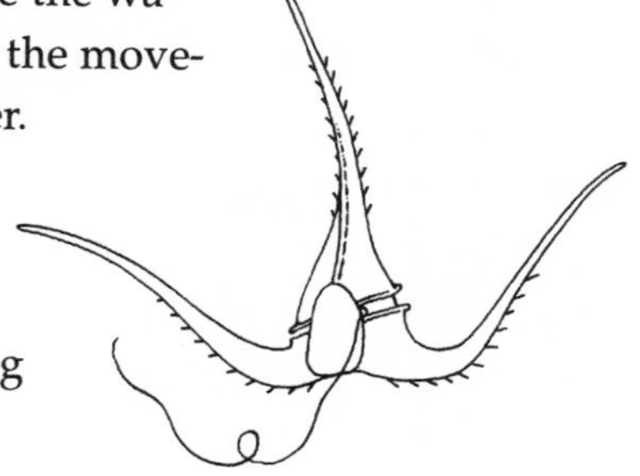

Ceratium *is one of the most common dinoflagellates. The flagellum that lies in the horizontal groove does not extend beyond the exoskeleton and is therefore not apparent.*

Experiments have shown that when actively bioluminescent dinoflagellates are mixed with copepods, the latter reduce their feeding rates on the former. The copepods do this by spending more time lunging and less time cruising, as if they are trying to escape from the dinoflagellates. Bioluminescence in dinoflagellates therefore seems to be a defense mechanism. Some researchers have suggested that the copepods' feeding is disturbed because they are startled by the light. Copepods do not have well-developed eyes like ours, but since the flashes look bright to us from a distance of several feet, they are probably quite intense from the copepod's perspective of only a few millimeters away. Other researchers think that the copepods' actions are an escape response, to get them away from other predators that might be attracted to the bright lights. Among the many enemies of copepods are fishes, which tend to hunt visually and are known to be attracted to bright lights; this is why fishermen often use bright lights and fluorescent lures. Dinoflagellates, being small, are usually swallowed whole. They remain alive inside their captor's gut for some little time, where they continue firing off their brilliant flashes of light. Most plankters have transparent bodies, and while this helps them to escape the attention of predators, it works against them when they have bioluminescent organisms in their gut! It is almost as if the doomed dinoflagellates are trying to bring about the demise of their captors by attracting fishes to the scene. Significantly, many transparent plankters, which live in tropical seas where there are intense bioluminescent displays, have darkly pigmented guts. These plankters include copepods and fish larvae, and the pigmentation helps hide the light flashes of engulfed dinoflagellates from prying eyes.

Gonyaulax is a small dinoflagellate, about 75 micrometers long. It often occurs in such dense concentrations that it colors the sea red, giving rise to the phenomenon of red tides. Like many other dinoflagellates, it is bioluminescent, but it is better known for causing paralytic shellfish poisoning. *Gonyaulax* produces a toxin that helps protect it against some predators, but its virulence varies among different groups of animals. Filter-feeders like mussels sometimes consume *Gonyaulax* in vast numbers, seemingly without any ill effects. But the toxin is extremely virulent to our own species, and if we eat infected mussels we can easily consume lethal amounts. The toxin attacks the nervous system, and the first symptoms, which develop within about fifteen minutes of eating the shellfish, are tingling sensations in the tongue and lips. This is followed by numbness, which can lead to death through paralysis of the respiratory system.

Many other microorganisms produce toxins, including, of course, bacteria. But when it comes to chemical warfare, plants are in a league of their own, as we will see in the next chapter.

9

The Interminable Struggle

*Low clouds drift across the rooftops like early morning mist. The scattering of stone houses that make up the village cling to the terraces carved in the land as if fearful of tumbling down the mountainside. The remote Himalayan kingdom of Bhutan is painted in the muted tones of earth and rock, but there are a few patches of green—a tree here, a low bush there. And outside the village are some small verdant plots of cultivated land used for growing buckwheat. But far more important to the economy of the village are the yaks (*Bos grunniens*). These shaggy-coated relatives of domesticated cattle are left to fend for themselves during the winter, but when the snow leaves the land they are herded up the mountain to their summer pastures.*

The yaks are used for milk, cheese, and meat, and are the mainstay of the village's economy. Consequently, when large numbers of yaks began dying of some unknown disease during the 1980s, the losses pressed heavily upon the villagers. In an attempt to determine the cause of the disease, a small team of veterinarians flew in from Australia.

The yaks are neither fed nor housed during the winter months and have to find what little grazing they can to sustain them, often by digging beneath the snow. They all lose weight and stop producing milk, and by the time spring returns to the Himalayas they are in poor condition. But they soon recover when grazing becomes plentiful, gaining weight and producing milk again. Yet some of the yaks fail to gain weight, even on the summer pastures, and either die during the summer or linger on into the next winter. The yak-herders are able to spot the sick animals without any difficulty. The earliest signs of disease are the appearance of small bald patches, with the characteristic lesions on the skin. These roughened areas are most common on the nose, legs, and back, and begin to weep, matting the surrounding hair and forming a scab. When the scab gets sloughed off, the area beneath is raw. Ulcers appear on the tongue, and the animals become emaciated. They are often unable to keep up with the rest of the herd as it moves on to fresh pastures. The less severely affected animals may survive into the winter, and their lesions may disappear completely. But they break out again during the spring, and the animals seldom survive another summer.

Autopsies performed on the dead yaks revealed that they all had abnormal livers. These were mottled with yellow, brown, and dark red areas, and the tissue was invaded with nodules. This type of liver damage is consistent with plant poisoning, so tissue samples were removed and tested chemically. The results confirmed the presence of a group of plant poisons called alkaloids (specifically pyrrolizidine alkaloids). Alkaloids are found in a wide variety of plant species, and it was necessary to find out which ones were causing the problem. Because of the remoteness of the area, there are no surveys of the flora, but *Senecio* is known to occur there. This genus of poisonous plants includes common groundsel (*Senecio vulgaris*) and ragwort (*S. jacobaea*), plants which cause similar cases of poisoning in livestock in many parts of the world. Domesticated animals are particularly prone to poisoning because they eat toxic plants that wild mammals usually avoid. A search for poisonous plants in the yak pastures of Bhutan revealed several species containing pyrrolizidine alkaloids. These included two species of *Senecio* (*S. raphanifolius* and *S. biligulatus*).

Alkaloids range in toxicity from stimulants like caffeine to deadly poisons like strychnine. They are just one of a number of groups of chemicals produced by plants and will be dealt with in some detail later in this chapter. There is much diversity among these groups of compounds, but all have one thing in common: they are harmful to animals. But why should plants want to wreak such havoc on animals? Think of it from the plant's perspective. Imagine being rooted in the ground, unable to move, and a yak comes along and starts nibbling on your limbs—a finger here, a toe there. What could you do to defend yourself? Some plants make themselves less amenable to large herbivores by having spines, like cactuses, or thorns, like brambles. Others, like trees, simply remove themselves from danger by growing tall. But defenses like these are not always effective and are of no avail against small herbivores like caterpillars and aphids. Plants have many other strategies for defending themselves against both small and large herbivores, but their most formidable weapons are chemical.

The plant kingdom's arsenal contains tens of thousands of chemicals. They range in subtlety from simple compounds, like cyanide, which kill, to analogues of hormones which can disrupt an animal's life cycle. Plants, like animals, also have to contend with invading microorganisms—viruses, bacteria, and fungi—and some of the antibiotics they manufacture to combat these organisms can be extracted and used in our own medicines. Notable among these is penicillin, extracted from the fungus *Penicillium,* often seen growing on the surface of moldy bread. The fungi themselves include some of the most common causes of accidental poisoning deaths in humans too. Among the most notorious of these is the destroying angel, or death cap (*Amanita phalloides*), a single mushroom of which can be fatal. Regardless of the importance of poisonous mushrooms from a human perspective, fungi are usually ignored by most other animals and will therefore not concern us here. Most of what I want to say about defense in plants pertains to the various compounds they manufacture, but before dipping into this chemical cornucopia, I will say something about some of their other strategies.

Plants that grow from the bottom, like grasses, are less vulnerable to the effects of herbivores than those that grow from buds, like trees and bushes. The golf course at the back of our house has its grass clipped every day, without any detriment to the plant. But if the surrounding trees received similar attention, they would probably all die. Not that the grasses acquiesce to being grazed upon without some resistance. Like other plants, their outer cell wall is constructed of cellulose, a tough material which has to be thoroughly ground up by the teeth before it can be digested properly. The more nutritious sap inside the cell contains small particles of silica—the same material as sand—which is so abrasive that they cause extensive tooth wear. If mammals that feed on grasses had ordinary teeth, they would soon be worn down and the individual would starve to death.[1] But herbivores overcame the problem by evolving high-crowned teeth that were long enough to last a lifetime. The change did not occur overnight. Grasses first appeared during the Miocene, about 20 million years ago, and the first mammals that grazed on them had low-crowned teeth. But over the course of the next several million years, there was a trend toward higher-crowned teeth—probably correlated with a trend toward larger body size and therefore longer life spans.[2] This trend reaches its highest development in the modern horse, whose high-crowned teeth are about 4 inches (10 cm) long.

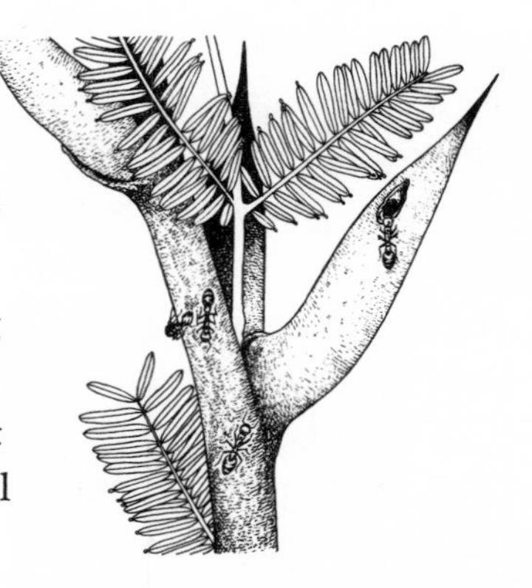

Certain ants live cooperatively on the acacia tree. They keep the plant free of other insects in return for food and shelter.

The efficacy of spines in discouraging large herbivores has often been questioned, and while a given plant's prickles may thwart some herbivores, they may be ineffectual against others. The fruits of the prickly pear cactus (*Opuntia*) could not look less appetizing, with their long, needle-sharp spines, but I saw land iguanas in Galápagos eating them as if there were no prickles at all. The acacia bush, or acacia tree (*Acacia*), found in arid regions in many parts of the world, is armed with vicious-looking thorns. These appear to discourage large mammalian herbivores, but they are no deterrent to small browsers, like insects. However, several species of *Acacia* have resolved this problem by entering into a symbiotic arrangement with a particular species of ant (*Pseudomyrmex ferruginea*). The ant does not harm the acacia, but the workers of this species are especially aggressive and drive off all other insects. If the ants are experimentally removed, the plant soon comes under attack from plant-eating insects. The ants' protective services are not offered for free, though. Aside from receiving shelter inside the hollow thorns, the ants are supplied with a particular kind of food. This takes the form of small nutritious swellings, called *Beltian bodies,* that sprout from the tips of the leaflets. The ants also receive nectar, so they are kept well fed. We humans find the sting of these ants particularly painful, with burning aftereffects which last for a long time. It seems likely that other mammals are similarly affected, so the ants probably deter large herbivores as well as small ones.

You may have noticed that some plants are hairy. The hairs may occur all over the plant, or they may be restricted to the leaves, usually to their undersides. I have just made a survey of the plants in my garden and can report that the following have hairy leaves: chrysanthemum (*Chrysanthemum*), honeysuckle (*Lonicera*), apple (*Malus*), pussy willow (*Salix*), and mock orange (*Philadelphus*). The mulberry (*Morus*) and geranium (*Pelagonium*) have some hairs, and these are restricted to the main ribs on the underside of the leaves. I should point out that none of the plants were obviously hairy, and I had to use a lens to see them. Why should plants be hairy? Most accounts propose that the hairs, or *trichomes,* have something to do with the plants' water balance, but there have been some suggestions that they are defensive. The idea that the trichomes might be protective dates back to the 1920s and 1930s when biologists working on crop pests like the potato leafhopper (*Empoasca fabae*) and the cotton leafhopper (*E. fascialis*) found that plants with the hairiest leaves had the fewest insects. Similar associations were found in the 1960s for the cereal leaf beetle (*Oulema melanopus*), a serious pest of grain crops in Europe and Asia that appeared in the United States in 1960. But why should being hairy protect a plant against such insects? The first phase of an insect infestation is when the female insect lays her eggs on the leaves of the host plant. If the leaf has a dense covering of hairs, the eggs cannot be attached directly to its surface and become attached to the hairs instead. The eggs, which are not protected by a shell, are therefore kept suspended in the air. This causes them to lose water more rapidly than if they were attached to the surface, because more of their water-permeable surface is exposed to the air. The water loss reduces hatching success, and less than 10 percent of the eggs laid on hairy leaves hatch, compared with 90 percent on hairless leaves.[3] The few insect larvae that do hatch out on hairy leaves do not survive very well because they have to eat their way through the trichomes, which are not very nutritive, before they can feed on the surface of the leaf.

Some trichomes are hooked, and these function to snare and even to impale insects. The French bean (*Phaseolus vulgaris*), for example, has hooked trichomes that catch in the claws of the feet of aphids. The ensnared insects starve to death, while others die from being impaled on the hooks. Similarly, the hooked trichomes that cover the entire surface of the passionflower vines (*Passiflora*) protect it against caterpillars. The caterpillars' soft bodies become impaled on the hooks, causing them to lose body fluids and die. Some species of *Passiflora* also have a strategy for discouraging female butterflies from laying their eggs on the leaves. These plants produce an outgrowth from the leaf that mimics the eggs of the butterfly. When a female lands on the leaf, she notices that another female has apparently been there already and laid eggs. She therefore finds somewhere else to lay her eggs, to reduce the competition for food when the caterpillars hatch out.

Anybody who grew up in England would be able to identify the stinging nettle (*Urtica*), because they have probably all had some painful contacts with the ubiquitous weed. The plant, whose leaf has a distinctive toothed margin, gets its name from the painful sting inflicted on brushing against it. The entire plant is covered with fine trichomes. They have sharp brittle points that readily break off, producing a cutting edge like the tip of a hypodermic needle. The broken trichomes readily puncture the skin, inoculating the victim with a fluid that produces a painful sting, lasting for an hour or more. It used to be thought that the active ingredients were formic and acetic acids, but the nettles manufacture a far more sophisticated potion that includes histamine, which causes a rash, and acetylcholine, which is involved in the transmission of nerve impulses. Most large herbivores, including horses, stay well clear of stinging nettles, but cattle seem to eat them with impunity. There are many other examples of trichomes that produce noxious secretions, one of the most ingenious being those of the wild potato (*Solanum*). The wild potato is prone to infestations of aphids and is densely covered with fine trichomes. When an aphid touches one of these, the hair ruptures, releasing a clear liquid which coats the aphid's legs. On contact with the air, the liquid solidifies into an insoluble black mass. As the aphid walks over the leaf, it inevitably trips other trichomes, and the accumulation of solid material on its legs soon immobilizes it. Since the aphid is unable to move, it cannot feed, and eventually starves to death. Such overt defense mechanisms as these are uncommon in plants, which tend toward more passive ways of protecting themselves.

Many plants, for example, have a thick, waxy cuticle covering their leaves, forming an effective barrier against many small organisms. Several examples can be found among indoor plants, including the jade plant (*Crassula argentea*), cacti, and the rubber plant (*Ficus elastica*). If rubber plants are damaged, their second line of defense is to release latex. Aside from containing toxins, the milky liquid solidifies when it dries. It is therefore effective against boring insects like leaf miners, which drill into plants, because it clogs up their feeding apparatus. Latex is also very bitter, and this may be a deterrent to herbivores. Chicory (*Cichorium intybus*), sometimes used as an additive to coffee, produces latex, making it taste bitter. Similar bitter compounds (called sesquiterpine lactones) occur in wild lettuce, deterring slugs from grazing on it. Domestic lettuce has been selectively cultivated to get rid of these unpalatable compounds, which explains why it is almost impossible to grow lettuce in the garden without attracting slugs. However, some varieties of lettuce, like the red ones, retain some of the bitter elements, giving them that extra "bite" that some people enjoy in their salads. This raises the question of the deterrent value of bitter-tasting compounds. While the bitter taste of certain plants may deter some herbivores, it is likely to be ineffective against others. Perhaps the bitter taste is merely incidental, and the real deterrent lies in the toxicity of the compounds that impart the bitterness. My own casual observations lead me to question the efficacy of bitterness as a deterrent, at least among some birds. In the fall and winter birds feed on a variety of berries, some of which look good to eat. But I have sampled lots of them over the years, and they almost always taste horribly bitter!

The chemicals considered in this chapter are compounds that do not appear to be involved in the plant's own metabolism. They are consequently called *secondary compounds.* There is no question that secondary compounds reduce the herbivorous activities of animals, but there has been much debate over whether they were primarily evolved for defense. We still have a great deal to learn about the biochemistry of plants, so it cannot be ruled out that some secondary compounds may be involved in the plant's own metabolism. However, if this were true, why are these compounds so diverse, and why do they differ from one species to another? At one time it was believed that secondary compounds were merely the waste products of plants, which just happened to be toxic or distasteful. This seems unlikely, though, because so many secondary compounds are removed from the leaves and retained by the plants prior to their leaves being shed in the fall. If they were waste products, the plant would have no need to retain them.

Regardless of their origins, the remarkable fact remains that plants manufacture tens of thousands of chemical compounds, many of which have profound effects on animals. Indeed, many of them, like those that influence the nervous system, can only affect animals because they alone have a nervous system. Some secondary compounds are so specific that they target a particular group of animals or even a single species. This is usually one of the insect species, reflecting the long period of geological time during which plants and insects have coexisted on Earth. Although plants produce some of the deadliest of poisons, and are responsible for a significant number of human and livestock deaths every year, they also synthesize substances that are of considerable benefit to us. This positive effect is entirely fortuitous, however, and we can be sure that each one of these compounds is detrimental to some other species. In general, this is true for all secondary compounds, as we will see, and substances that deter one species often act as an attractant for others. The cocklebur (*Xanthium*), for example, is an herbaceous weed that produces a compound to protect it against seed-eating moths. But the same compound appears to attract a seed-eating fly. It is probably because of this variation in effects among different animal species that plants manufacture such a wide array of defensive chemicals. They fall into several different categories, but we will primarily be concerned with two types, the tannins and the alkaloids.

That quintessential element of summertime in North America, the cottage, typically nestles among the trees on the edge of a lake. The runoff from the land percolates through the woods before reaching the lake, which is why its waters are usually discolored. The brown stain is attributed to the tannins leached into the water from the fallen leaves and dead trees of the forest. Tannins are also responsible for the color and mild bitter taste of tea, and for the slight astringency of red wine, derived from the grape skins. Although the strong bitterness of many tannins may deter some herbivores, their physiological effects are more damaging. A characteristic feature of tannins is their readiness to bind with proteins, and for this reason they have been used for centuries in tanning leather. Many plants, including oaks (*Quercus*), have leaves that are rich in tannins, but these do not give complete protection from herbivores, neither large nor small. The effect of ingesting tannins is that they bind with the proteolytic enzymes in the gut—the enzymes that digest proteins—interfering with digestion. The tannins also bind with other proteins in the gut, causing additional disruption of the digestive process. It used to be thought that protein binding was the only reaction of tannins, but it is now known that they are able to bind with other molecules, too, including starch and cellulose. In the best-case scenario, the herbivore suffers what amounts to a bad case of indigestion, but at the other extreme the results are fatal. Much probably depends on the amounts of tannin consumed relative to the size of the animal. Domestic animals—cattle, sheep, and goats—readily browse on oak leaves, especially fresh buds and immature leaves. The toxicity of oak is low, and if it does not exceed about 50 percent of their diet, no harm is done. But these levels are often exceeded, especially in free-ranging cattle in the American Southwest, where deaths from oak poisoning can be high. Symptoms appear within about a week and rapidly worsen. The animals become anorexic, gut movements cease, causing constipation, and they become excessively thirsty. If the animals do not receive treatment, they start passing blood from the gut, the heartbeats become rapid but weaker, and the animals die. Postmortem examination reveals a characteristic degradation of the kidneys, diagnostic of tannin poisoning.

During the summer months, meadow voles (*Microtus pennsylvanicus*) and prairie voles (*M. ochrogaster*) living in the Midwest of the United States eat large quantities of a plant (*Lespedeza*) containing high levels of tannins. The plant comprises 28 percent of their diet, but laboratory trials have shown that the voles eat very little of it when more palatable food is available. To find out the effect of the tannins on growth, extracts were made from the plant, mixed with artificial food, and fed to young voles of both species. The prairie voles were unaffected by the tannins during the three-week trial period, but the growth of the meadow voles was retarded. These investigations show how animals can tolerate limited quantities of tannins with no apparent ill effects, and also how closely related species can react differently to the same toxins.

The alkaloids, which include the earliest medicines used by humans, are among the most interesting of the secondary compounds. They were once thought to be solely of plant origin, but some alkaloids are now known to be manufactured by animals, though these are of no interest to us here. It has been estimated that about 20 percent of plants manufacture alkaloids, and about six thousand alkaloids (or compounds with alkaloid properties) have so far been recognized. I won't go into their chemical structure, except to say that they all possess at least one hexagonal ring of carbon atoms (called an aromatic ring, or benzene ring) with a nitrogen atom. They all appear to have some physiological effects on animals, often on the central nervous system, and many of them are highly toxic. One of the most familiar alkaloids, consumed in varying quantities by most of us every day, is caffeine. Caffeine occurs naturally in the coffee plant (*Coffea arabica*) and tea plant (*Thea sinensis*), where it may serve to discourage herbivorous insects. Whatever its effects may be on insects, its physiological action on our own species is well known. Aside from inducing a state of wakefulness, caffeine increases the pressure of the blood in the kidneys, stimulating the production of urine (diuresis). Caffeine can also induce irregularities in the heartbeat (arrhythmia). On the other side of the coin, caffeine can help relieve headaches and is a component of several preparatory brands of aspirin-based analgesics such as Anadin.

Another alkaloid that figures prominently in human affairs is nicotine. We usually associate this alkaloid with cultivated tobacco (*Nicotiana tabacum*) and the dependency it engenders for smoking. However, there are several wild species, too, all producing varying amounts of the alkaloid, and the effects that nicotine has on smokers is quite incidental to its primary role as a powerful toxin to discourage herbivory. The leaves of all species are unpalatable, but livestock, given the chance, will feed upon them, often with fatal results. Experiments with one of the species (*Nicotiana trigonophylla*) have shown that the minimum dose required to produce symptoms in livestock is about 2 percent of the animal's body weight. The symptoms, which can appear in as little as fifteen minutes, primarily involve the nervous system. They include shivering, sweating, muscle spasms, vomiting, diarrhea, staggering, weakness, and eventual collapse. The heart may start beating violently, but the pulse soon becomes weak and rapid. Breathing difficulties sometimes appear, due to paralysis of the respiratory muscles. Death may occur within minutes or within several days of ingesting the plants. Horses have died after eating harvested tobacco leaves, and pigs have succumbed after getting loose in a planted field of tobacco. There is also a report of a family being poisoned, one of them fatally, after eating wild tobacco leaves that they had boiled up as a vegetable.

The tobacco hornworm (*Manduca sexta*) is a caterpillar that feeds upon the leaves of tobacco plants, both the cultivated and the wild species. The caterpillars are able to tolerate the nicotine they ingest, provided the levels in the leaves are not excessively high. However, the tobacco plant, like many other plant species, is able to increase the amount of alkaloids it manufactures in response to being damaged. The total leaf alkaloids can increase by as much as four times following injury. So what effect would a rise in alkaloid levels have on the caterpillars feeding on the plants? To find out, freshly hatched caterpillars were raised on wild tobacco plants (*Nicotiana sylvestris*), some of which were undamaged and had normal nicotine levels, others having been damaged and therefore having elevated levels. The caterpillars on the undamaged plants developed normally, but the others grew far more slowly. They gained only about one-third as much weight as the other caterpillars, and this was attributed to their lowered appetites. Depression of the appetite is one of the effects nicotine has on our own species, which is one of the reasons for the gain in weight that usually accompanies successful attempts to quit smoking. The stunted caterpillars also had reduced survival rates. Although nicotine does not give tobacco plants complete protection from being browsed upon by insects, they

Deadly nightshade has plump black berries that some children find irresistible. All parts of the plant are poisonous.

are able to respond to heavy damage by depressing their attackers' appetites and by reducing their chances of survival.

The toxic component of the foxglove is digitalis, one of the effects of which is to slow the heart.

When I was young, I was given dire warnings of two poisonous plants, deadly nightshade (*Atropa belladonna*) and the foxglove (*Digitalis purpurea*). In each case the toxic component has an effect on the heart and has been extracted and used as a drug for many years. The toxin in deadly nightshade is atropine, which speeds up the heart, while that of the foxglove is digitalis, which slows the heart down and makes it beat more strongly. All parts of the plants are toxic, but the incidence of poisonings, either of humans or of livestock, is usually rare. The single exception is deadly nightshade poisoning of children, the reason being that the plant has glossy black berries that children find very inviting. Deaths have been reported for children who have eaten as few as three berries. Aside from increasing the heart rate, deadly nightshade poisoning causes wide dilation of the pupils and elevated body temperatures, sometimes leading to mental confusion and unconsciousness. The symptoms of foxglove poisoning include trembling, vomiting, diarrhea, abdominal pains, and drowsiness. Mammalian herbivores appear to avoid both plants, and the potent alkaloids presumably discourage herbivorous insects too.

Plants and insects have coexisted for over 300 million years, maintaining a delicate balance between their respective needs. Most plants are dependent upon insects for cross-pollination, while many insects depend on plants for food. This mutualism has resulted in some remarkable instances of coevolution, where evolutionary change in one brings about an evolutionary change in the other, which in turn causes change in the first, and so on. If plants had not evolved an elaborate system of secondary compounds to deter overgrazing by insects (and other herbivores), they would have been eaten out of existence. And if the herbivorous insects had not evolved metabolic pathways for detoxifying many of these chemical weapons, they would have been poisoned into oblivion. This, in turn, would have deprived plants of their pollinating services, leading to their own demise. Plants and insects therefore maintain a delicate balancing act, neither side winning, neither side losing to the other. And it is because of the fine-tuning between plants and animals that certain species of insects are usually associated with a particular species of host plant. If you scan the pages of a butterfly book, for example, you will see that most species lay their eggs on one or a few particular plant species, to which their caterpillars are adapted. The red admiral (*Vanessa atalanta*), for example, typically lays its eggs on stinging nettles, whereas the painted lady (*V. cardui*) usually chooses thistles.

Insects appear to have a general detoxification mechanism for dealing with the wide array of chemicals they encounter, rather than having multiple mechanisms for targeting individual toxins. This enables them to detoxify some compounds that they have never encountered before. It also permits significant improvements to evolve in their detoxification abilities through relatively small changes in their genes. This helps explain how resistance to certain synthetic insecticides, like DDT, has evolved so rapidly. These compounds were introduced at about the time of the Second World War, but it has not taken many years for insects to become resistant to them. Unfortunately the "higher" animals, the vertebrates, have not been able to evolve such resistance, and the heavy losses seen among animals at the top of the food pyramid, like raptors, are a direct consequence of these poisons. Since the 1960s, populations of most species of raptors have plummeted worldwide, bringing some close to the point of extinction.

The yaks in Bhutan were poisoned by the class of alkaloids called pyrrolizidine alkaloids (PAs). These toxins are responsible for significant livestock deaths in many parts of the world. It seems that mammals are particularly vulnerable to PAs, but insects are able to detoxify them. Many insects even actively seek out and feed on plants that contain high levels of the toxins. As a result, the alkaloids, which are intensely bitter, are taken up by their tissues, making them distasteful to birds and other predators. Two species of butterflies (*Danaus chrysippus* and *D. gilippus*) that are closely related to the monarch (*D. plexippus*), for example, acquire PAs by visiting certain plants that have high levels of the toxin. In each case it is the male that collects the toxin, thereby becoming unpalatable. Like the monarch butterfly, which is also unpalatable, these butterflies are strikingly colored, and this serves as a warning of their unpalatability to their avian predators. Such warning colors are described as *aposematic*. Aside from achieving a level of immunity from attack, the male uses some of its sequestered alkaloid to manufacture a pheromone, used to attract females.

During copulation the male transfers a large proportion of the PA he has collected—well over half—to the female via his semen. The female, in turn, transfers most of this to her eggs, giving them protection from being eaten. The eggs are probably the most vulnerable part of the butterfly's life cycle, and experiments on other species have shown that eggs containing PA are usually avoided by predators. The male's alkaloid gift to the female is therefore a significant contribution to the survival of the next generation. The male's ability to attract a mate is determined by his ability to sequester PA and manufacture the sex pheromone. The more PA the male has collected, the more pheromone he will be able to release to attract the female. It is therefore possible that the male can tell the female just how good he is at sequestering PA, and therefore how good a provider he will be for the protection of the eggs. In evolutionary terms, the male may be capable of communicating his fitness to the female.

The monarch butterfly, which looks just like its two relatives, also takes up toxins from plants to make itself unpalatable. However, it is the larvae rather than the adults that do the sequestering, by feeding on milkweed (*Asclepias*). The toxin in milkweed is not an alkaloid but a compound that has a chemical structure similar to that of sugars, and which is called a glycoside. The toxin accumulates in the caterpillars as they voraciously feed on the plants and is passed on to the adult when the caterpillar undergoes its transformation into a butterfly. Since both the male and female butterflies acquire the toxin, each one makes a contribution to the protection of the egg. Incidentally, the viceroy butterfly, which does not sequester toxins and which is therefore palatable, mimics the monarch butterfly. The viceroy butterfly therefore acquires protection from birds simply because it looks like the bitter-tasting monarch butterfly. Probably all unpalatable butterflies acquire their bad taste by taking up toxins from plants.

Moths also sequester toxins from plants, and one of those that has been studied in some detail belongs to the genus *Creatonotos*. One of the remarkable features of these moths is that the male has a huge fernlike organ at the end of its abdomen which is used for discharging sex pheromone into the air. The size of the structure is directly related to the quantity of PAs the male ac-

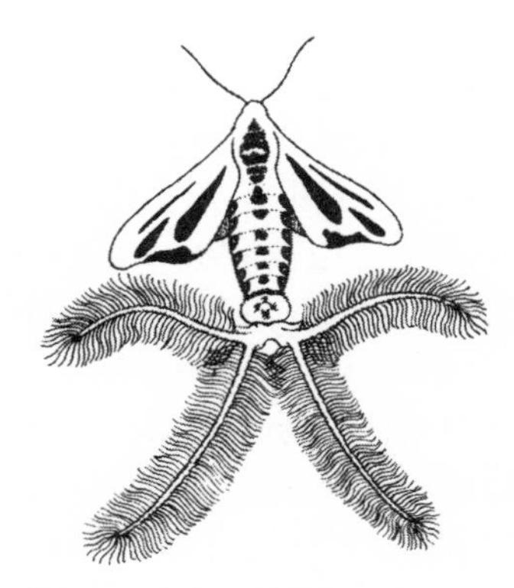

The male moth of the genus Creatonotos *is remarkable for its large scent organ, the size of which increases with the quantities of alkaloids the moth extracts from its plant food.*

cumulated in its body when it was a caterpillar. If caterpillars are reared on plants with low PA levels, the pheromone organ barely develops. But those reared on plants with high alkaloid levels develop huge organs, bigger than the rest of the body.

The mutually beneficial relationships that have evolved between certain plants and insects can be just as intricate as some of our own business relationships. Take, for example, the case of the weed, the aphid, the ant, and the moth. The weed is the ragwort (*Senecio jacobaea*), which we have already encountered in connection with PA poisoning in yaks. The levels of PA in the ragwort vary between plants by a factor of about ten. An aphid, appropriately named *Aphis jacobaeae*, lives on the ragwort, choosing plants with the lowest PA levels. Like all other aphids, it makes a living by piercing the plant with its hypodermic mouthparts and sucking up the plant sap. If the ragwort had an opinion on the matter, it would probably admit to not liking it a lot. But there are far worse things in life, like being eaten alive by caterpillars of a moth named *Tyria jacobaeae*. These caterpillars attack the ragwort in the early summer, over a period of about six weeks, during which time their insatiable appetites can completely defoliate a plant. Not even ragworts with high PA levels can deter the caterpillars. However, plants with resident aphids are left completely untouched. This is not because the aphids are a threat to the caterpillars—these small, soft-bodied insects are vulnerable to attack themselves. The reason has to do with the company the aphids keep: an ant named *Lasius niger*. The aphids supply the ants with honeydew, and the ants, in return, jealously protect them. No other insects are allowed on the plant, including the hungry caterpillars. So the plant enjoys protection from the caterpillars because its PA levels are low enough to attract aphids, which produce honeydew that is not too bitter with alkaloids for the ants!

Not all aphids avoid plants with high alkaloid levels. The aphid *Macrosiphon albifrons*, which lives on lupins (*Lupinus*), selects plants with high concentrations of alkaloids. It does this to gain protection from a beetle (*Carabus problematicus*) that preys upon it. Selective feeding experiments have shown that aphids reared on lupins with low alkaloid levels are eaten in preference to those that live on plants rich in alkaloids. Sometimes beetles do make mistakes, though, and occasionally consume an aphid that has high alkaloid levels. These beetles are literally put on their backs by the toxins and do not recover for about forty-eight hours.

The damage aphids inflict on plants may be mild compared with that done by caterpillars, but a heavy infection can be a serious drain on the plant's resources. Many plants therefore discourage aphids, and these include the wild potato (*Solanum berthaultii*). As we have seen, this species has a novel way of dealing with aphids which involves the trichomes (hairs) that cover the leaves and stem of the plant. There are two kinds of trichomes, slender ones that end in a small ball, and short ones ending in a larger subdivided ball. When aphids, or any other insect, brush against the short trichomes, the balls rupture and release a sticky fluid that quickly sets, trapping the intruder. Rupture of the tall trichomes also releases a sticky droplet, but this one contains a stockpile of secondary compounds. Included among them is one of the main components of aphid alarm pheromone. When this is released into the air, it serves as a warning to other aphids to stay away, reducing the risk of a reinvasion. The cultivated potato plant (*Solanum tuberosum*) lacks the short trichomes, and the other ones probably do not release the alarm hormone. The plant is therefore more vulnerable to aphid infestations. However, since the two species readily hybridize, it should be possible to transfer the protective trichomes of the wild species to the cultivar.

The fact that a plant can produce an insect pheromone to dispel aphids is even more impressive to me than the bola spider's synthesis of a pheromone to attract moths (see the discussion in chapter 6). But this is not an isolated instance, because many plants synthesize insect hormones or compounds that mimic them. These synthetic products imitate two classes of insect hormones: juvenile hormones, and molting hormones, both of which control molting. *Molting* is a necessary part of the life cycle of insects and other arthropods, because of the nature of their skeletons. Unlike vertebrates, which have their skeletons on the inside, arthropods have theirs on the outside, like a suit of armor. An exoskeleton has the advantage of offering protection as well as support, but suffers the disadvantage of not allowing for growth. An arthropod overcomes the problem by molting several times during its life, shedding its old exoskeleton during the process and forming another one. The new exoskeleton, which forms beneath the old one, is soft at first and stays that way for several hours. This allows the animal to puff itself up before the new exoskeleton hardens, giving it room to grow. The molts are initiated by the release of the molting hormone. If juvenile hormone is also present, the insect molts into another juvenile stage, but if absent, the individual undergoes a metamorphosis into the adult, as when a caterpillar is transformed into a butterfly. By synthesizing analogues of these two hormones, plants are able to disrupt the normal life cycle of their unwanted guests, often resulting in their sterility and death.

Sometimes plants manipulate insects to serve their own purposes. A good example of this is provided by the cucumber plant (*Cucumis*). When attacked by herbivorous mites, it secretes a secondary compound (a terpenoid) which acts as an attractant to predatory mites. When the predators arrive on the plant, they begin feeding on the herbivorous mites, ridding the plant of the problem.

I have only touched on some of the chemical defenses of plants, having said nothing about the strongly aromatic species like sage (*Artemisia*), creosote (*Larrea*), mesquite (*Prosopis*), and pine (*Pinus*); nor of plants with dangerous levels of oxalates in their leaves, like rhubarb (*Rheum*), sorrel (*Oxalis*), and beet (*Beta*). There are also many plants that manufacture cyanide, including white clover (*Trifolium*), hydrangea (*Hydrangea*), flax (*Linum*), and, surprisingly, the following fruit trees: apricot, cherry, chokecherry, peach, plum, sloe (all of which belong to the genus *Prunus*), and apple (*Malus*). I hasten to point out that the only poisonous parts of these plants are the kernels of the pits, or stones.

Some plants concentrate toxic elements, like selenium, which they take up from the soil. Whether this is a defensive strategy is not certain, but the results of eating some of these plants can be fatal. Plants that concentrate selenium include poison vetches (*Astragalus*), aster (*Aster*), tansy aster (*Machaeranthera*), saltbush (*Atriplex*), and ironweed (*Sideranthus*). There are also plants like the Venus flytrap (*Dionaea muscipula*) and sundew (*Drosera rotundifolia*) that devour small insects.

Plants, for all their outward appearance of benign tranquillity, merit our caution and respect. All the death and suffering they have meted out to humankind over the millennia is nought compared with the benefits we have derived from their chemical armory. Aside from such useful products as latex (for rubber), turpentine (as a solvent and a base for paints), and creosote (for preserving wood), plants gave us all our early drugs. These include quinine, morphine, codeine, colchicine, atropine, ephedrine, plus many others. And, starting with the discovery of penicillin in 1928, fungi provided us with our first antibiotics.

Every human culture has its own traditional medicines for dealing with particular maladies, and although some of the prescriptions are no more effective than the proverbial chicken soup, others contain factors of known pharmacological activity. Leaves of the plant *Aspilia*, for example, widely used in African native medicine for a range of disorders, contain thiarubrine-A, a potent agent against bacteria, fungi, and parasitic flatworms. The plant is used not only by the human population of Tanzania but also by the wild chimpanzees (*Pan troglodytes*). The chimpanzees collect the leaves on a regular basis, usually first thing in the morning. The leaf is not chewed but is held inside the mouth and massaged against the cheeks by the tongue. It has been suggested that this technique evolved in chimpanzees to enhance the oral uptake of the drug, because it is inactivated by the acidity of the stomach. We use similar methods for taking drugs that are sensitive to stomach acids. The nitroglycerin that is prescribed for angina, for example, is taken in the form of capsules that are dissolved beneath the tongue. There is also a report of an anorexic and obviously unwell chimpanzee that was seen sucking bitter juices from the pith of a certain tree (*Vernonia amygdalina*). Extracts from this tree are widely used in Africa for the treatment of parasitic infections and for gastrointestinal disorders. Presumably, the sick primate had actively sought out the unpleasant-tasting plant because of its medicinal properties.

The use of drugs by nonhuman animals may not be confined to chimpanzees. Baboons (*Papio*), for example, have been seen eating the fruit of *Balanites aegyptiaca*. The fruit may simply have been eaten as a source of food. However, it is possible that the baboons were exploiting its medicinal properties because the fruit contains high concentrations of diosgenin, a steroid drug that is effective against the larval stages of parasitic flatworms (flukes).

Over 90 percent of the biological compounds on the Earth's surface are produced by plants. Many of these compounds have pharmacological properties and are already being used as drugs. Many more still have to be discovered. The need to discover new drugs has become a matter of great urgency lately because of the appearance of so many new strains of pathogens that have evolved resistance to our current drugs. Old enemies that we thought we had beaten, like tuberculosis, are on the rampage again, and strep throat, once considered a minor irritant, has become a potential killer. Then there are the new agents of destruction, like the AIDS and ebola viruses. The tropical rain forests of the world, the last great bastions of biodiversity, are a promising source of new drugs, and collecting trips have been made to the tropics to gather new plants for this purpose. Our salvation from the epidemics that must surely come may well lie among the plants, but *Homo sapiens*, the wise man, is destroying the rain forests, and all the other natural habitats, faster than we can study them.

Epilogue: In an English Country Garden

The weathered red brick, aglow in the afternoon sun, radiates warmth like a fine cognac. The wall has stood at the bottom of the garden since the reign of Queen Anne. More than two hundred years of sun, wind, and frost have softened the stolid features into a crumbling craggy homeliness, where mosses and ivy-leaved toadflax grow in profusion.

A bumblebee, legs pantalooned in pollen, emerges from a foxglove's bell and drones off to her next floral embrace. A thrush sings a pastorale to the heavens atop a cherry tree, speckled breast sparkling in the sun.

Several generations of gardeners have tended this rural half acre over the years. And, in spite of all the changes the world has seen, they have managed to preserve something of its original character. An ornamental pond still stands at the bottom of the garden, but the fountain has gone, and sticklebacks and frogs have supplanted goldfish and carp. The present owners prefer a natural look to manicured lawns; wild flowers are encouraged and weeds tolerated. And if the garden looks a little overgrown and unkept, so much the better for the birds and squirrels, and for the fox that has taken up residence in the village. But the owners are not averse to some judicious husbandry, and the master of the house, sleeves rolled, brow beaded, tills the heavy dark soil of his wife's vegetable garden.

Distant church bells peel out the hour—a pious reminder that today is the sabbath. The resonant strains of the last chime drift across the warm summer air like dandelion down. The thrush quits its lofty perch and swoops to the freshly turned soil. Its glistening beak stabs into the earth, emerging the next instant holding the end of a worm. The thrush tugs and pulls with all its might, stretching the end of the reluctant worm like a piece of elastic. But the worm is too big and too fat to be budged from its underground haven. Undaunted, the determined thrush secures a new hold closer to the ground and starts heaving again. Still the worm stays fast in its burrow. Now the thrush pecks and shakes at the recalcitrant worm as if in blind fury and succeeds in breaking off a writhing tidbit. The bleeding stump disappears beneath the soil as the thrush devours its gritty prize.

The roses should be good again this year—a myriad of tight buds attest to this—but some of the plants are heavily infested with aphids. The plump green insects crawl over the leaves on their spindly legs like overweight guests at a banquet. They plunge their hypodermics into the plants' living tissue and drink freely of the sap. A ladybird, all crimson red and shiny black spots, alights on a rose. It makes its way up the stem toward a leaf, encountering an aphid along the way. The aphid is grasped, ripped apart, and consumed. The voracious beetle continues up the stem. When it reaches the leaf, it begins systematically working its way through the assembled aphids—a miniature siege engine, tearing and devouring until its appetite is satiated.

Ladybird, ladybird, fly away home.
Your house is on fire and your children have gone.

Unblinking eyes stare out from the pond. Only the head lies above the surface; the rest of the body is hidden among the weeds. The frog has remained frozen in the water for the last twenty minutes. There is not even a ripple on the surface to reveal its presence.

A small tortoiseshell butterfly flits among the flowers, resplendent in its orange-and-brown livery, piped in blue. It leaves the blossoms and flutters toward the pond, settling on a lily pad and folding back its wings. How drab and brown it looks now, with its bright colors hidden from view. After a few minutes the butterfly takes off again, but it has scarcely become airborne when its delicate airframe is rocked by a violent and immobilizing blow. The next instant it is engulfed by a cavernous mouth rimmed by sharp teeth. The jaws snap closed, then open again. Using its clumsy hands, the frog attempts to swallow the prey. It is a laborious business, but jaws and hands work away, and eventually the butterfly's broken body can be fed into the frog's slimy gullet. With one final gulp and a blinking of eyes, the butterfly disappears.

Death lurks at the bottom of the pond too. Here the predator is a dragonfly larva, and its intended victim is a tadpole. Adult and larval dragonflies alike are rapacious carnivores. The adult feeds on flying insects, which it takes on the wing, whereas its offspring dine on aquatic fare. And while the adult's life is numbered in weeks, the larva takes upward of two years to mature.

The dragonfly larva, almost as long as a thumb and reminiscent of an earwig, clambers over a clump of submerged waterweed. Its enormous black eyes catch sight of the tadpole, but it is too far away, so the larva has to get closer. Beneath its chin, and stretching partway back along its body, is a hinged apparatus, armed with hooked jaws. The folded appendage can be shot out in an instant to impale its prey, but it has to get within range first. The larva inches forward.

The tadpole, its membranous tail undulating incessantly, grazes on a sprig of Canadian pond weed. It attacks the bright green leaflets voraciously, scalloping the margins with its tiny jaws. The plant has already been nibbled by other tadpoles and is not at its best.

Without warning, a pair of sharp fangs pierce the tadpole's body and whisk it through the water to the black staring eyes. The dragonfly holds the tadpole in its deadly embrace and injects digestive enzymes into the soft body. The corrosive fluids begin their work, and the tadpole thrashes its life away in the water.

The back door opens, and the woman of the house steps out into the sunshine, her young son in tow. She asks her husband to keep an eye on the toddler while she finishes cooking the Sunday roast. The tiller of soil smiles and waves, and carries on with his labors.

The world is an enchanted place for a small child, with so many wonders to behold. The little boy sees a ladybird fly into a web and watches in rapt awe as the spider rushes out to attend to its struggling captive. An inquisitive hand reaches out for the web, but the sticky threads adhere to his fingers. He pulls back his hand, and the web, the spider, and the ladybird disappear by magic. He coos and chuckles and moves on.

Now the little explorer is on his haunches beside a flower bed. He is looking at a plump caterpillar, busily munching on a nasturtium leaf. Waves of contraction ripple down the length of the green-and-yellow body as the caterpillar inches its way along the leaf. These plants must be very tasty, because the caterpillar has already eaten a lot of the leaf. And, to the little boy's extreme delight, he sees that a second creepy-crawly is having a piggyback ride on the caterpillar. The second insect, balancing on slender legs like a ballerina, has its slim abdomen arched down, caressing the caterpillar's back.

The hitchhiker is an ichneumon wasp and is in the act of depositing her eggs inside the caterpillar's body. When the eggs hatch, they will live inside the caterpillar, feeding on its tissues and growing as the caterpillar grows. The unwilling host will be little affected by their brooding presence, but when it is time to change into a butterfly, the larvae will erupt from its body, so killing it.

The attention span of two-year-olds is short, and the little boy moves on to other wonders. Here is a nice big plant taller than himself. There are no caterpillars to be seen, but just look at those shiny black berries. Each plump berry sits on a platter of tiny green leaves, like a miniature plum on a plate. He reaches out, picks several berries, and pops them into his mouth.

The church bell tolls.

Notes

Chapter 1. Nature, Red in Tooth and Claw

1. Cats' eyes are adapted for night vision. The light-sensitive cells of the retina predominate in rods, the ones that respond to light and dark rather than to color (attributed to the cone cells). The retina is also backed by a well-developed reflective layer that bounces light back through the rods again, increasing sensitivity even further. We lack this reflective layer, and the "red eye" result of flash photography is due to the bright light reflecting from the rich blood supply of the retina. A flashlight is too weak to achieve the same effect, even at night, but if you shone one into a cat's eyes they would flash back at you like beacons.
2. The cheetah, like many other carnivores, bounds along when it runs, and the muscular flexion and extension of its vertebral column greatly increases the length of its stride, thereby increasing its speed. Cheetahs have especially flexible and slender backs, which, coupled with their long slender legs, gives them their remarkable speed.
3. Hunting success, as used here and elsewhere in the book, is the number of times prey are caught, expressed as a percentage of the number of times an attempt is initiated by the predator to capture a potential prey animal. Suppose a person wanted to know the success rate of single lions hunting zebras. He or she might spend several months in Africa watching single lions hunting zebras. Suppose, further, that this person watched, and saw the outcome of, a total of 214 attempts. This total would, of course, include aborted attempts, as when a stalking lion is spotted by the prey and therefore the lion does not even bother to chase it. But it would not include any attacks that lions made on nonprey, such as when a lion chases a hyena away from its kill. Suppose that of all 214 attempts, the single lions managed to kill zebras only 37 times. The hunting success would be 37 divided by 214 times 100, or 17 percent.

4. Cheetahs might more accurately be described as pursuit predators, because of their long charges and because of the long distances they often cover during hunting forays. Some researchers refer to them as such, while others describe them as ambush predators.

Chapter 2. In Cold Blood

1. Endotherms do not always maintain a constant body temperature day and night. Most birds usually drop their temperatures by one or two degrees at night; hummingbirds drop their temperatures down close to ambient (outside) levels. Roosting bats also lower their body temperatures, as do hibernating mammals. Our own body temperature falls slightly at night, but this is by less than one degree (F).
2. The fifth surviving reptilian group, the sphenodontids, comprises a single species, the tuatara, a lizardlike reptile found only in New Zealand.
3. John Ruben 1995, personal communication. He described how he himself was once chased by a mamba in an enclosure at a zoo when he was making some observations. The snake pursued him over a distance of several yards and struck at him, narrowly missing his chest.
4. This estimate of annual snakebite deaths is given by Habernehl 1981. There are between six thousand and seven thousand cases of venomous snakebites each year in the United States, but few fatalities occur. See Minton 1987.
5. The rates of chemical processes increase with temperature, doubling for every rise of ten degrees (C). This is called the Q_{10} effect. Non-endothermic animals are influenced by the Q_{10} effect, the rates of their body processes increasing with temperature.
6. See Auffenberg 1987, p. 302.

Chapter 3. Death at Sea

1. The distinction is made throughout this book between degrees Fahrenheit (°F) and Fahrenheit degrees (F°), the former being a particular temperature, the latter being a temperature difference. For example, 40°F is a particular temperature, which happens to be 8 F° above freezing point. The same is true for temperatures on the Celsius scale. Thus 105 C° is 105 divisions on the Celsius scale, whereas 105°C is a particular temperature which happens to be 5 C° above the boiling point of water.
2. William John Dakin, D.Sc, a zoologist who was skeptical of the story of the killer whales of Twofold Bay, visited the area in 1932. This was just two years after the death of "Old Tom," the last killer whale of the pod. He interviewed witnesses, looked at the whaling station and its equipment, and examined the skeletal remains of Old Tom. Remarkably, he found a diary, written in Twofold Bay in 1843, in which similarly cooperative behavior of killer whales was reported, almost a century before. Dakin's findings convinced him of the truth of the "Killers of Eden."
3. There is an historical precedent for naming these structures by the rather fanciful name of "monkey lips." See Cranford, Amundin, and Norris 1996, for further reading.

Chapter 4. Avoiding Attention

1. An ion is an electrically charged particle. Ions are formed when the atoms or molecules of certain elements are dissolved in water. Common salt, for example, has the molecular formula NaCl, and its molecule is made up of an atom of sodium (Na) joined to an atom of chlorine (Cl). When dissolved in water, the molecule breaks up into sodium ions, which are positively charged, and chlorine ions, which are negatively charged.
2. Personal communication from Peter J. Herring 1996.
3. Shrimps take on their familiar restaurant pink when they are boiled.
4. The body length referred to here does not include the tail. The chameleon species investigated, *Chamaeleo oustaleti,* had a body length (minus the tail) of 9 inches (22–23 cm). For further details, see Wainwright, Kraklau, and Bennett 1991.
5. This extract, from an undated book by Stewart E. White, *The Rediscovered Country,* published in London, was cited by Cott 1940, p. 94.

Chapter 5. Air Strike

1. A female hen harrier weighs about 1 pound (0.45 kg); a female peregrine falcon, up to about 1.6 pounds (0.76 kg).
2. Strictly speaking, the term mass should be used rather than *weight.* Mass is the quantity of matter and is independent of gravity, whereas weight varies with gravity. The mass of a stone may be 2.2 pounds (1 kg) and would be exactly the same on the Moon. The same stone would also weigh 2.2 pounds on Earth, but its weight on the Moon would only be about 0.22 pounds because the gravity of the Moon is about one-tenth that of Earth. In everyday usage, *mass* and *weight* are used interchangeably, and I will not make the distinction between them in the text of this book.
3. The reason wing loading increases with increasing body length is that area (the numerator in the expression for wing loading) increases with the square of the linear dimension, whereas the mass (denominator) increases with the cube. For further reading on the subject, see C. McGowan, *Diatoms to Dinosaurs: The Size and Scale of Living Things* (Washington, D.C.: Island Press, 1994).
4. For further reading on this high-lift device, and for more information on soaring flight, see the reference in the previous note.
5. Small animals consume more food than large ones, relative to their body mass, because their metabolic rates are so much higher. For an explanation, see the reference in note 3.
6. The fine processes on the dorsal surfaces of the wing feathers of owls are extensions of the *barbules,* which are side branches from the barbs. The prominent fringe along the leading edge of the first wing feather is formed by extensions of the barbs. Similarly, the fringe along the trailing edges of the primary and secondary feathers is an extension of the barbs.
7. The focal length of a lens is defined as the point on the principal axis through which all rays that were parallel to the principal axis pass after refraction. Simply stated, if you hold a lens up to a wall opposite a window and move it back and forth, a point will be reached when a small image of the window is brought into sharp fo-

cus on the wall. The distance between the wall and the lens is the focal length of the lens.

8. An oscilloscope is an instrument that displays sounds visually on a television tube.

Chapter 6. The Scorpion's Sting

1. For more information on size and scale, see C. McGowan, *Diatoms to Dinosaurs: The Size and Scale of Living Things* (Washington, D.C.: Island Press, 1994).
2. Data from P. Matthews, *The Guinness Book of Records,* 1994 (New York: Bantam Books, 1994).
3. Professor John Moore of Northern Colorado University informs me that certain red inks used in ballpoint pens, including Bic pens, contain a chemical similar to that of scent trails. If you draw a line using one of these pens, termites, and possibly ants, will follow it.
4. Termite species with the snapping jaws belong to the genera *Capritermes, Neocapritermes,* and *Pericapritermes.* Some peoples use these insects to suture wounds.
5. Mild steel has a tensile strength of $4.3–4.9 \times 10^8$ newtons per square meter, or pascals (Pa), compared with 10×10^8 Pa for spider silk. Values taken from the following sources, respectively: G. W. C. Kay and T. H. Laby, *Table of Physical and Chemical Constants* (London: Longman, 1973); R. McN. Alexander, *The Invertebrates* (Cambridge: Cambridge University Press, 1979).
6. The life spans for male and female redback spiders pertain to animals kept in captivity, but there is no reason to believe that similar disparities between the sexes do not occur in the wild.

Chapter 7. Saurian Warriors

1. I use the term *dinosaur* in its widest sense, which is exclusive of birds. However, I acknowledge that birds, which are classified as theropods, really are dinosaurs. Reasons for rejecting *carnosaur* are given in Holtz 1994.
2. I discussed the question of dinosaurian metabolism and thermal strategies in some detail elsewhere. For more information, see McGowan 1979, 1991.
3. Dromaeosaurs may have had high metabolic rates like modern birds. However, John Ruben (personal communication) reports that there is no evidence of respiratory turbinate bones in dromaeosaurs. These bones, located in the nasal passage, are always correlated with endothermy, and their absence weighs heavily against dinosaurian endothermy. For more information, see Ruben 1995.
4. The thermal strategy of maintaining fairly constant body temperatures by virtue of an animal's thermal inertia is referred to as *inertial homeothermy.*
5. The closest living relatives of dinosaurs are birds. Indeed, the relationship is so close that birds and dinosaurs are now classified within the same group, the Dinosauria. The extinct animals that are popularly known as dinosaurs are accordingly referred to as nonavian dinosaurs, to distinguish them from the other dinosaurs, which are feathered and popularly known as birds. I use the terms *dinosaur* and *bird* in the popular sense. *Reptile* is also used, in the popular sense, for the ectothermic tetrapods, the snakes, lizards, crocodiles, turtles, and the tuatara (*Sphenodon*), a lizardlike relic from the Mesozoic Era.

6. A Boeing 727-100 has a maximum gross weight of 169,500 pounds (75.6 tons), and the estimated weight of *Brachiosaurus* is 78 tons. Data from Taylor 1975 and Colbert 1962.
7. Most of the mounted skeleton of the Berlin specimen of *Brachiosaurus* is made of plaster. The original bones are in the storage area, thereby being more accessible to researchers.
8. A new carnivorous dinosaur has recently been discovered in Argentina that may be larger than *Tyrannosaurus.* This Cretaceous dinosaur, named *Gigantosaurus,* is more robustly built than *Tyrannosaurus* and is estimated to have been about 41 feet (12.5 m) long and to have weighed 6–8 tons. For further information, see Coria and Salgado 1995.
9. Ichthyosaur remains had been found before Mary Anning's first specimen, and reported in the literature, but these accounts seem to have had little impact. See, for example, Delair 1969, Young 1819.
10. The Crystal Palace, a magnificent wrought iron and glass edifice built to house the Great Exhibition of 1851, was originally constructed in London's Hyde Park. It was dismantled in 1854 and relocated in Sydenham, now part of the greater London area. Quite by chance I met one of the firemen who attended the fire that destroyed the building in 1936. In his opinion the building could have been saved.
11. An imperfect skeleton, found in 1803 and examined by Sir Everard Home in 1818, disappeared soon afterward. See Delair 1969 for further information.
12. Although this specimen, which is in the museum at Stuttgart, may have died during natural birth, it may well have been a postmortem birth. Such events occur in nature after the death of a pregnant female. As putrefaction proceeds, the mounting gas pressure inside the abdomen forces the dead fetus out of the adult's body cavity. For further information, see McGowan 1991 and the references therein.

Chapter 8. Warfare in Miniature

1. Animals with strong swimming powers, like most fishes and all the cetaceans, are collectively referred to as the *nekton,* from the Greek word for swimming.
2. The freshwater genera studied were the copepods *Cyclops* and *Epischura,* and the cladocerans *Bosmina* and *Chydorus.* For more information, see Kerfoot 1978.

Chapter 9. The Interminable Struggle

1. Much of the tooth wear from feeding on grasses is attributed to the abrasive particles from the ground that are ingested with the graze. See Janis and Fortelius 1988.
2. Larger animals tend to live longer than smaller ones. The ancestors of the modern horse were about the size of large dogs and probably lived less than half as long as modern horses. For further reading on the relationship between size and longevity, see C. McGowan, *Diatoms to Dinosaurs: The Size and Scale of Living Things* (Washington, D.C.: Island Press, 1994).
3. Data for hatching success pertain to the cereal leaf beetle. For more information, see Levin 1973.

Bibliography

Prologue

Abrahams, P. A. 1986. Adaptive responses of predators to prey and prey to predators: The failure of the arms-race analogy. *Evolution* 40:1229–47.

McGowan, C. 1994. *Diatoms to dinosaurs: The size and scale of living things.* Washington, D.C.: Island Press.

Meech, L. D. 1966. *The wolves of Isle Royale.* Fauna of the National Parks of the United States, Fauna Series 7. Washington, D.C.: Department of the Interior.

Chapter 1. Nature, Red in Tooth and Claw

Abrahams, P. A. 1986. Adaptive responses of predators to prey and prey to predators: The failure of the arms-race analogy. *Evolution* 40:1229–47.

Akersten, W. A. 1985. Canine function in *Smilodon* (Mammalia; Felidae; Machairodontinae). *Contributions in Science,* Natural History Museum of Los Angeles County, 356:1–22.

Alexander, R. McN. 1993. Legs and locomotion of Carnivora. In *Mammals as predators,* edited by N. Dunstone and M. L. Gorman, 65:1–13. Oxford: Symposium of the Zoological Society of London.

Bakker, R. T. 1983. The deer flies, the wolf pursues: Incongruencies in predator-prey evolution. In *Coevolution,* edited by D. J. Futuyma and M. Slatkin, pp. 350–82. Sunderland, Mass.: Sinauer.

Caro, T. M., and C. D. Fitzgibbon. 1992. Large carnivores and their prey: The quick and the dead. In *Natural enemies: The population biology of predators, parasites, and diseases,* edited by M. J. Crawley, pp. 117–42. Oxford: Blackwell.

Case, R., and J. Stevenson. 1991. Observation of barren-ground grizzly bear, *Ursus arctos,* predation on muskoxen, *Ovibovis moschatus*, in the Northwest Territories. *Canadian Field Naturalist* 105:105–6.

Danilov, P. I. 1983. The brown bear (*Ursus arctos* L) as a predator in the European taiga. *Acta Zoologica Fennica* 174:159–60.

Estes, R. D. 1991. *The behavior guide to African mammals.* Berkeley: University of California Press.

Ewer, R. F. 1973. *The Carnivores.* London: Weidenfeld and Nicolson.

Fuller, T. K., and P. W. Kat. 1993. Hunting success of African wild dogs in southwestern Kenya. *Journal of Mammalogy* 74:464–67.

Futuyma, D. J., and M. Slatkin. 1983. Introduction. In *Coevolution,* edited by D. J. Futuyma and M. Slatkin, pp. 1–13. Sunderland, Mass.: Sinauer.

Gittleman, J. L. 1989. Carnivore group living: Comparative trends. In *Carnivore behavior, ecology, and evolution,* edited by J. L. Gittleman, pp. 183–207. Ithaca: Cornell University Press.

Glickman, S. E., L. G. Frank, P. Licht, T. Yalcinkaya, P. K. Siiteri, and J. Davidson. 1992. Sexual differentiation of the female spotted hyena. *Annals of the New York Academy of Sciences* 662:135–59.

Harvey, P. H., and J. L. Gittleman. 1992. Correlates of carnivory: Approaches and answers. In *Natural enemies: The population biology of predators, parasites, and diseases,* edited by M. J. Crawley, pp. 26–39. Oxford: Blackwell.

Janis, C. 1976. The evolutionary strategy of the Equidae and the origins of rumen and cecal digestion. *Evolution* 30:757–74.

———. 1986. Evolution of horns and related structures in hoofed mammals. *Discovery* 19:9–17.

———. 1989. A climatic explanation for patterns of evolutionary diversity in ungulate mammals. *Palaeontology* 32:463–81.

———. 1990. Correlation of reproductive and digestive strategies in the evolution of cranial appendages. In *Horns, pronghorns, and antlers: Evolution, morphology, physiology, and social significance,* edited by G. A. Bubenik and A. B. Bubenik, pp. 114–33. New York: Springer-Verlag.

———. 1994. The sabertooth's repeat performance. *Natural History* 4:78–83.

Janis, C., and P. B. Wilhelm. 1993. Were there mammalian pursuit predators in the Tertiary? Dances with wolf Avatars. *Journal of Mammalian Evolution* 1:103–25.

Laurenson, M. K. 1994. High juvenile mortality in cheetahs (*Acinonyx jubatus*) and its consequences for maternal care. *Journal of Zoology* 234:387–403.

Lindstedt, S. L., J. F. Hokanson, D. J. Wells, S. D. Swain, H. Hoppeler, and V. Navarro. 1991. Running energetics in the pronghorn antelope. *Nature* 353:748–50.

Lundrigan, B. 1996. Morphology of horns and fighting behavior in the family Bovidae. *Journal of Mammology* 77:462–75.

Matthews, N. E., and W. F. Porter. 1988. Black bear predation of white-tailed deer neonates in the central Adirondacks. *Canadian Journal of Zoology* 66:1241–42.

Mills, M. G. L., and H. C. Biggs. 1993. Prey apportionment and related ecological relationships between large carnivores in Kruger National Park. In *Mammals as predators,* edited by N. Dunstone and M. L. Gorman, 65:253–268. Oxford: Symposium of the Zoological Society of London.

Mysterud, I. 1975. Sheep killing and feeding behaviour of the brown bear (*Ursus arctos*) in Trysil, south Norway 1973. *Norwegian Journal of Zoology* 23:243–60.

Polis, G. A. 1991. Complex trophic interactions in deserts: An empirical critique of food-web theory. *American Naturalist* 138:123–55.

Schaller, G. B. 1972. *The Serengeti lion*. Chicago: University of Chicago Press.

Scheel, D. 1993. Profitability, encounter rates, and prey choice of African lions. *Behavioral Ecology* 4:90–97.

Seip, D. R. 1992. Factors limiting woodland caribou populations and their interrelationships with wolves and moose in southeastern British Columbia. *Canadian Journal of Zoology* 70:1494–1503.

Stander, P. E., and S. D. Albon. 1993. Hunting success of lions in a semi-arid environment. In *Mammals as predators,* edited by N. Dunstone and M. L. Gorman, 65:127–43. Oxford: Symposium of the Zoological Society of London.

Van Valkenburgh, B. 1985. Locomotor diversity within past and present guilds of large predatory mammals. *Paleobiology* 11:406–28.

———. 1988. Incidence of tooth breakage among large predatory mammals. *American Naturalist* 131:291–302.

———. 1989. Carnivore dental adaptations and diet: A study of trophic diversity within guilds. In *Carnivore behavior, ecology, and evolution.* edited by J. L. Gittleman, pp. 410–36. Ithaca: Cornell University Press.

———. 1996. Feeding behavior in free-ranging, large African carnivores. *Journal of Mammology* 77:240–54.

Van Valkenburgh, B., and K-P. Koepfli. 1993. Cranial and dental adaptations to predation in canids. In *Mammals as predators,* edited by N. Dunstone and M. L. Gorman, 65:15–37. Oxford: Symposium of the Zoological Society of London.

Chapter 2. In Cold Blood

Arnold, S. J. 1993. Foraging theory and prey-size–predator-size relations in snakes. In *Snakes: Ecology and behavior,* edited by R. A. Seigel and J. T. Collins, pp. 117–64. New York: McGraw-Hill.

Auffenberg, W. 1981. *The behavioral ecology of the Komodo monitor.* Gainsville: University Presses of Florida.

———. 1987. Social and feeding behavior in *Varanus komodoensis*. In *Behavior and neurobiology of lizards,* edited by N. Greenberg and P. D. MacLean, pp. 301–31. Rockville, Md.: National Institute of Mental Health.

Bellairs, A. 1969. *The life of reptiles.* London: Weidenfeld and Nicolson.

Bennett, A. F. 1982. The energetics of reptilian activity. In *Biology of the Reptilia,* edited by C. Gans and F. H. Pough, pp. 155–99. London: Academic Press.

Bennett, A. F., R. S. Seymour, D. F. Bradford, and G. J. W. Webb. 1985. Mass-dependence of anaerobic metabolism and acid-base disturbance during activity in the salt-water crocodile, *Crocodylus porosus. Journal of Experimental Biology* 118:161–71.

Brodie, E. D., III, and E. D. Brodie, Jr. 1990. Tetrodotoxin resistance in garter snakes: An evolutionary response of predators to dangerous prey. *Evolution* 44:651–59.

Brodie, E. D., Jr., and D. R. Formanowicz, Jr. 1980. Palatability and antipredator behavior of the treefrog *Hyla versicolor* to the shrew *Blarina brevicauda. Journal of Herpetology* 15:235–36.

Brodie, E. D., Jr., R. A. Nussbaum, and M. DiGiovanni. 1984. Antipredator adaptations of Asian salamanders (Salamandridae). *Herpetologica* 40:56–68.

Cock Buning, T. de. 1983. Thermal sensitivity as a specialization for prey capture and feeding in snakes. *American Zoologist* 23:363–75.

Cott, H. B. 1961. Scientific results of an inquiry into the ecology and economic status of the Nile crocodile (*Crocodilus niloticus*) in Uganda and Northern Rhodesia. *Transactions of the Zoological Society of London* 29:211–356.

Cundall, D. 1987. Functional morphology. In *Snakes: Ecology and evolutionary biology,* edited by R. A. Seigel, J. T. Collins, and S. S. Novak, pp. 106–40. New York: Macmillan.

Daltry, J. C., W. Wüster, and R. S. Thorpe. 1996. Diet and snake venom evolution. *Nature* 379:537–40.

Elliott, W. B. 1978. Chemistry and immunology of reptilian venoms. In *Biology of the Reptilia,* edited by C. Gans and K. A. Gans, pp. 163–436. London: Academic Press.

Fenton, M. B., and L. E. Licht. 1990. Why rattle snake? *Journal of Herpetology* 24:274–79.

Garland, T., Jr. 1988. Genetic basis of activity metabolism. I. Inheritance of speed, stamina, and antipredator displays in the garter snake *Thamnophis sirtalis. Evolution* 42:335–50.

Habernehl, G. G. 1981. *Venomous animals and their toxins.* Berlin: Springer-Verlag.

Kardong, K. V. 1975. Prey capture in the cottonmouth snake (*Agkistrodon piscivorus*). *Journal of Herpetology* 9:169–75.

———. 1979. "Protovipers" and the evolution of snake fangs. *Evolution* 33:433–43.

———. 1980. Evolutionary patterns in advanced snakes. *American Zoologist* 20:269–82.

———. 1982. Comparative study of changes in prey capture behavior in the cottonmouth (*Agkistrodon piscivorus*) and Egyptian cobra (*Naja haje*). *Copeia* 1982:337–43.

———. 1986. The predatory strike of the rattlesnake: When things go wrong. *Copeia* 1986:816–20.

Kardong, K. V., and P. A. Lavin-Murcio. 1993. Venom delivery of snakes as high-pressure and low-pressure systems. *Copeia* 1993:644–50.

Kardong, K. V., and S. P. Mackessy. 1991. The strike behavior of a congenitally blind rattlesnake. *Journal of Herpetology* 25:208–11.

Klauber, L. M. 1956. *Rattlesnakes.* Berkeley: University of California Press.

Latifi, M. 1978. Commercial production of anti-snakebite serum (antivenin). In *Biology of the Reptilia* 8, edited by C. Gans and K. A. Gans, pp. 561–88. London: Academic Press.

Lillywhite, H. B. 1987. Temperature, energetics, and physiological ecology. In *Snakes: Ecology and evolutionary biology,* edited by R. A. Seigel, J. T. Collins, and S. S. Novak, pp. 422–77. New York: Macmillan.

Lillywhite, H. B., and R. W. Henderson. 1993. Behavioral and functional ecology of arboreal snakes. In *Snakes: Ecology and behavior,* edited by R. A. Seigel and J. T. Collins, pp. 1–48. New York: McGraw-Hill.

Mebs, D. 1978. Pharmacology of reptilian venoms. In *Biology of the Reptilia* 8, edited by C. Gans and K. A. Gans, pp. 437–560. London: Academic Press.

Middendorf, G. A., III., and W. C. Sherbrooke. 1992. Canid elicitation of blood-squirting in a horned lizard (*Phrynosoma cornutum*). *Copeia* 1992:512–27.

Minton, S. A. 1987. Poisonous snakes and snakebite in the U.S.: A brief review. *Northwest Science* 61:130–37.

Mushinsky, H. R. 1987. Foraging ecology. In *Snakes: Ecology and evolutionary biology,* edited by R. A. Seigel, J. T. Collins, and S. S. Novak, pp. 302–34. New York: Macmillan.

Nedospasov, A. A., and A. V. Cherkasov. 1993. Radioactivity of snake venom. *Nature* 361:409.

Pooley, A. C., and C. Gans. 1976. The Nile crocodile. *Scientific American* 234:114–24.

Rasmussen, S., B. Young, and H. Krimm. 1995. On the "spitting" behavior in cobras (Serpentes: Elapidae). *Journal of Zoology, London,* 237: 27–35.

Rowe, M. P., and D. H. Owings. 1978. The meaning of the sound of rattling by rattlesnakes to Californian ground squirrels. *Behavior* 66:252–67.

Ruben, J. A. 1976. Aerobic and anaerobic metabolism during activity in snakes. *Journal of Comparative Physiology* 109:147–57.

———. 1977. Some correlates of cranial and cervical morphology with predatory modes in snakes. *Journal of Morphology* 152:89–99.

———. 1979. Blood physiology during activity in the snakes *Masticophis flagellum* (Colubridae) and *Crotalus viridis* (Crotalidae). *Comparative Biochemistry and Physiology* 64:577–80.

———. 1989. Activity physiology and evolution of the vertebrate skeleton. *American Zoologist* 29:195–203.

Ruben, J. A., and C. Geddes. 1983. Some morphological correlates of striking snakes. *Copeia* 1983:221–25.

Savitsky, A. H. 1980. The role of venom delivery strategies in snake evolution. *Evolution* 34:1194–1204.

Schwenk, K. 1994. Why snakes have forked tongues. *Science* 263:1573–77.

Scudder, K. M. 1983. Effect of environmental odors on strike-induced chemosensory searching by rattlesnakes. *Copeia* 1983:519–22.

Secor, S. M., and J. Diamond. 1995. Adaptive responses to feeding in Burmese pythons: Pay before pumping. *Journal of Experimental Biology* 198:1313–25.

Shine, R. 1990. Function and evolution of the frill of the frillneck lizard, *Chlamydosaurus kingii* (Sauria: Agamidae). *Biological Journal of the Linnean Society* 40:11–20.

Wüster, W., and R. S. Thorpe. 1992. Dentitional phenomena in cobras revisited: Spitting and fang structure in the Asiatic species of *Naja* (Serpentes: Elapidae). *Herpetologica* 48:424–34.

Chapter 3. Death at Sea

Armstrong, B. 1981. *Sable Island.* Toronto: Doubleday.

Arnold, P. W. 1972. Predation on harbour porpoise, *Phocoena phocoena,* by a white shark, *Carcharodon carcharias. Journal of the Fisheries Research Board of Canada* 29:1213–14.

Baldridge, A. 1972. Killer whales attack and eat a gray whale. *Journal of Mammalogy* 53:898–900.

Block, B. A., and D. Booth. 1992. Direct measurement of swimming speeds and depth of blue marlin. *Journal of Experimental Biology* 166:267–84.

Brodie, P., and B. Beck. 1983. Predation by sharks on the grey seal (*Halichoerus grypus*) in eastern Canada. *Canadian Journal of Fisheries and Aquatic Sciences* 40:267–71.

Caldwell, D. K., and D. H. Brown. 1964. Tooth wear as a correlate of described feeding behavior by the killer whale, with notes on a captive specimen. *Bulletin of the Southern California Academy of Sciences* 63:128–40.

Campagno, L. J. V. 1984. Sharks of the world. FAO Species Catalogue. Food and Agriculture Organization of the United Nations, FAO Fisheries Synopsis, no. 125, vol. 4, pt. 1.

Campbell, L. 1974. *Sable Island, fatal and fertile crescent*. Windsor, Nova Scotia: Lancelot Press.

Carey, F. G. 1982. A brain heater in the swordfish. *Science* 216:1327–29.

Carey, F. G., J. W. Kanwisher, O. Brazier, G. Gabrielson, J. G. Casey, and H. L. Pratt. 1982. Temperature activities of a white shark, *Carcharodon carcharias*. *Copeia* 1982:254–60.

Carey, F. G., and B. H. Robinson. 1981. Daily patterns in the activities of swordfish, *Xiphias gladius*, observed by acoustic telemetry. *Fisheries Bulletin* 79:277–92.

Carey, F. G., and J. V. Scharold. 1990. Movement of blue sharks (*Prionace glauca*) in depth and course. *Marine Biology* 106:329–42.

Condy, P. R., R. J. Van Aarde, and M. N. Bester. 1978. The seasonal occurrence and behaviour of killer whales, *Orcinus orca*, at Marion Island. *Journal of Zoology* 184:449–64.

Cranford, T. W., M. Amundin, and K. S. Norris. 1996. Functional morphology and homology in the odontocete nasal complex: Implications for sound generation. *Journal of Morphology* 228:223–85.

Dakin, W. J. 1934. *Whalemen adventurers*. Sydney: Angus and Robertson. (Reprinted by Sirius Books in 1963.)

Day, L. R., and H. D. Fisher. 1954. Notes on the great white shark, *Carcharodon carcharias*, in Canadian Atlantic waters. *Copeia* 1954:295–96.

Dolphin, W. F. 1987. Observations of humpback whale, *Megaptera novaeangliae*, killer whale, *Orcinus orca*, interactions in Alaska: Comparison with terrestrial predator-prey relationships. *Canadian Field Naturalist* 101:70–75.

Engaña, A. C., and J. E. McCosker. 1984. Attacks on divers by white sharks in Chile. *California Fish and Game* 70:173–79.

Frost, P. G. H., P. D. Shaughnessy, A. Semmelink, M. Sketch, and W. R. Siegfried. 1975. The response of jackass penguins to killer whale vocalisations. *South African Journal of Science* 71:157–58.

Hancock, D. 1965. Killer whales kill and eat a minke whale. *Journal of Mammalogy* 46:341–42.

Klimley, A. P. 1994. The predatory behavior of the white shark. *American Scientist* 82:122–33.

Le Boeuf, B. J. 1982. White shark predation on pinnipeds in California coastal waters. *Fishery Bulletin* 80:891–95.

Lopez, J. C., and D. Lopez. 1985. Killer whales (*Orcinus orca*) of Patagonia, and their behavior of intentional stranding while hunting nearshore. *Journal of Mammalogy* 66:181–83.

MacGintie, G. E., and N. MacGintie. 1968. *Natural history of marine animals*. New York.: McGraw-Hill.

Martin, A. 1993. Hammerhead shark origins. *Nature* 364:494.

Mead, T. 1961. *Killers of Eden*. Pymble, New South Wales: Angus and Robertson. (Reprinted edition.)

Morejohn, G. V. 1968. A killer whale—gray whale encounter. *Journal of Mammalogy* 49:327–28.

Norris, K. S., and B. Møhl. 1983. Can odontocetes debilitate prey with sound? *American Naturalist* 122:85–104.

Norris, K. S., and J. H. Prescott. 1961. *Observations on Pacific cetaceans of Californian and Mexican waters*. Berkeley: University of California Press.

Northridge, S., and J. Beddington. 1992. Marine mammals. In *Natural enemies: The popu-*

lation biology of predators, parasites, and diseases, edited by M. J. Crawley, pp. 188–204. Oxford: Blackwell.

Odlum, G. C. 1948. An instance of killer whales feeding on ducks. *Canadian Field Naturalist* 62:42.

Powlik, J. J. 1995. On the geometry and mechanics of tooth position in the white shark, *Carcharodon carcharias. Journal of Morphology* 226:277–88.

Pratt, H. L., and J. G. Casey. 1982. Observations on large white sharks, *Carcharodon carcharias,* off Long Island, New York. *Fishery Bulletin* 80:153–56.

Randall, J. E. 1973. Size of the great white shark (Carcharodon). *Science* 181:169–70.

Rice, D. W. 1968. Stomach contents and feeding behavior of killer whales in the eastern North Pacific. *Norsk Hvalfangsttidende* 57:35–38.

Schevill, W. E., and W. A. Watkins. 1966. Sound structure and directionality in *Orcinus* (killer whale). 1966. *Woods Hole Oceanographic Institution* 1787:71–76.

Scott, W. B., and M. G. Scott. 1988. *Atlantic fishes of Canada.* Canadian Bulletin of Fisheries and Aquatic Sciences 219. (Joint publication with University of Toronto Press.)

Skud, E. 1962. Measurements of a white shark, *Carcharodon carcharias,* taken in Maine waters. *Copeia* 1962:659–61.

Smith, T. G., D. B. Siniff, R. Reichle, and S. Stone. 1981. Coordinated behavior of killer whales, *Orcinus orca,* hunting a crabeater seal, *Lobodon carcinophagus. Canadian Journal of Zoology* 59:1185–89.

Strong, W. R., F. F. Snelson, and S. M. Gruber. 1990. Hammerhead shark predation: An observation of prey handling by *Sphyrna mokarran. Copeia* 1990:836–40.

Tarpy, C. 1979. Killer whale attack! *National Geographic* 155:542–45.

Tricas, T. C., and J. E. McCosker. 1984. Predatory behavior of the white shark (*Carcharodon carcharias*), with notes on its biology. *Proceedings of the California Academy of Sciences* 43:221–38.

Voisin, J. F. 1972. Notes on the behaviour of the killer whale, *Orcinus orca* (L.). *Norwegian Journal of Zoology* 20:93–96.

Chapter 4. Avoiding Attention

Barrows, W. B. 1913. Concealing action of the bittern (*Botaurus lentiginosus*). *Ibis* 30:187–90.

Cott, H. B. 1940. *Adaptive coloration in animals.* London: Methuen.

Dumbacher, J. P., B. M. Beehler, T. F. Spande, H. M. Garraffo, and J. W. Daly. 1992. Homobatrachotoxin in the genus *Pitohui:* Chemical defense in birds? *Science* 258:799–801.

Edmunds, M., and J. Grayson. 1989. Camouflage and selective predation in caterpillars of the poplar and eyed hawkmoths (*Laothoe populi* and *Smerinthus ocellata*). *Biological Journal of the Linnean Society* 42:467–80.

Eisner, T., K. Hicks, M. Eisner, and D. S. Robson. 1978. "Wolf-in-sheep's-clothing" strategy of a predaceous insect larva. *Science* 199:790–93.

Endler, J. A. 1978. A predator's view of animal color patterns. *Evolutionary Biology* 11:319–64.

Fishlyn, D. A., and D. W. Phillips. 1980. Chemical camouflaging and behavioral defenses against a predatory seastar by three species of gastropods from the surfgrass *Phyllospadix* community. *Biological Bulletin* 158:34–48.

Forester, A. J. 1979. The association between the sponge *Halichondria panicea* (Pallas) and scallop *Chlamys varia* (L.): A commensal-protective mutualism. *Journal of Experimental Marine Biology and Ecology* 36:1–10.

Gilbert, L. E. 1983. Coevolution and mimicry. In *Coevolution,* edited by D. J. Futuyma and M. Slatkin, pp. 263–81. Sunderland, Mass.: Sinauer.

Götmark, F. 1987. White underparts in gulls function as hunting camouflage. *Animal Behavior* 35:1786–92.

———. 1994. Does a novel bright colour patch increase or decrease predation? Red wings reduce predation risk in European blackbirds. *Proceedings of the Royal Society of London* 256:83–87.

Götmark, F., and A. Hohlfält. 1996. Bright male plumage and predation risk in passerine birds: Are males easier to detect than females? *Oikos* 74:475–84.

Greene, E. 1989. A diet-induced developmental polymorphism in a caterpillar. *Science* 243:643–46.

Guilford, T. 1992. Predator psychology and the evolution of prey coloration. In *Natural enemies: The population biology of predators, parasites, and diseases,* edited by M. J. Crawley, pp. 377–94. Oxford: Blackwell.

Hanlon, R. T., and R. F. Hixon. 1980. Body patterning and field observations of *Octopus burryi* Voss, 1950. *Bulletin of Marine Science* 30:749–55.

Herring, P. J. 1977. Bioluminescence of marine organisms. *Nature* 267:788–93.

Kilar, J. A., and R. M. Lou. 1986. The subtleties of camouflage and dietary preference of the decorator crab, *Microphrys bicornutus* Latreille (Decapoda: Brachyura). *Journal of Experimental Marine Biology and Ecology* 101:143–60.

McFall-Ngai, M., and J. G. Morin. 1991. Camouflage by disruptive illumination in leiognathids, a family of shallow-water bioluminescent fishes. *Journal of Experimental Biology* 156:119–37.

Mills, D. 1993. *Aquarium fish.* London: Dorling Kindersley.

Moritz, R. F. A., W. H. Kirchner, and R. M. Crewe. 1991. Chemical camouflage on the death's head hawkmoth (*Acherontia atropos* L.) in honeybee colonies. *Naturwissenschaften* 78:179–82.

Neudecker, S. 1989. Eye camouflage and false eyespots: Chaetodontid responses to predators. *Environmental Biology of Fishes* 25:143–57.

Reilly, J. C. 1970. Project Yehudi. A study of camouflage by illumination, 1935–1945. *Journal of the American Aviation Historical Society* 15:255–62.

Silberglied, R. E., A. Aiello, and D. M. Windsor. 1980. Disruptive coloration in butterflies: Lack of support in *Anartia fatima. Science* 209:617–19.

Stuckenberg, B. 1981. The striping of zebras: A new look at old data. *Antenna* 5:145–48.

Wainwright, P. C., D. M. Kraklau, and A. F. Bennett. 1991. Kinematics of tongue projection in *Chamaeleo oustaleti. Journal of Experimental Biology* 159:109–33.

Wicksten, M. K. 1978. The exterior decorator. *Sea Frontiers* 24:277–80.

———. 1983. Camouflage in marine invertebrates. *Oceanography and Marine Biology* 21:177–93.

Chapter 5. Air Strike

Alerstam, T. 1987. Radar observations of the stoop of the peregrine falcon *Falco peregrinus* and the goshawk *Accipiter gentilis. Ibis* 129:267–73.

Austin, O. L. 1962. *Birds of the world.* London: Hamlyn.

Barton, N. W. H., and D. C. Houston. 1994. Morphological adaption of the digestive tract in relation to feeding ecology of raptors. *Journal of Zoology, London* 232:133–50.

Bates, D. L., and M. B. Fenton. 1990. Aposematism or startle? Predators learn their responses to the defenses of prey. *Canadian Journal of Zoology* 68:49–52.

Brown, L., and D. Amadon. 1968. *Eagles, hawks, and falcons of the world.* 2 vols. New York: McGraw-Hill.

Dekker, D. 1980. Hunting success rates, foraging habits, and prey selection of peregrine falcons migrating through central Alberta. *Canadian Field Naturalist* 94:371–82.

Fenton, M. B. 1983. *Just bats.* Toronto: University of Toronto Press.

Fenton, M. B., D. Audet, M. K. Obrist, and J. Rydell. 1995. Signal strength, timing, and self-deafening: The evolution of echolocation in bats. *Paleobiology* 21:229–42.

Fenton, M. B., and J. H. Fullard. 1981. Moth hearing and the feeding strategies of bats. *American Scientist* 69:266–75.

Fox, R., S. W. Lehmkuhle, and D. H. Westendorf. 1976. Falcon visual acuity. *Science* 192:263–65.

Fullard, J. H., J. A. Simmons, and P. A. Saillant. 1994. Jamming bat echolocation: The dogbane tiger moth *Cynia tenera* times its clicks to the terminal attack calls of the big brown bat *Eptesicus fuscus*. *Journal of Experimental Biology* 194:285–98.

Harris, J. T. 1979. *The peregrine falcon in Greenland.* Columbia: University of Missouri Press.

Hill, J. E., and J. D. Smith. 1984. *Bats: A natural history.* Austin: University of Texas Press.

Houston, D. C. 1986. Scavenging efficiency of turkey vultures in tropical forest. *Condor* 88:318–23.

———. 1988. Digestive efficiency and hunting behaviour in cats, dogs, and vultures. *Journal of Zoology, London* 216:603–5.

Jaksić, F. M., and J. H. Carothers. 1985. Ecological, morphological, and bioenergetic correlates of hunting mode in hawks and owls. *Ornis Scandinavica* 16:165–72.

Martin, G. R. 1977. Absolute visual threshold and scotopic spectral sensitivity in the tawny owl *Strix aluco*. *Nature* 268:636–38.

Mendelsohn, J. M., A. C. Kemp, H. C. Biggs, R. Biggs, and C. J. Brown. 1989. Wing areas, wing loadings, and wing spans of 66 species of African raptors. *Ostrich* 60:35–42.

Newton, I. 1979. *Population ecology of raptors.* Vermillion, S. Dak.: Buteo Books.

———. 1992. Birds of prey. In *Natural enemies: The population biology of predators, parasites, and diseases,* edited by M. J. Crawley, pp. 143–62. Oxford: Blackwell.

Nicholls, M. K., and R. Clark, eds. 1993. *Biology and conservation of small falcons.* London: Hawk and Owl Trust.

Norberg, R. A. 1978. Skull asymmetry, ear structure and function, and auditory localization in Tengmalm's owl, *Aegolius funereus* (Linné). *Philosophical Transactions of the Royal Society of London,* B 282:325–410.

Obrist, M. K., M. B. Fenton, J. L. Eger, and P. A. Schlegel. 1993. What ears do for bats: A comparative study of pinna sound pressure transformations in Chiroptera. *Journal of Experimental Biology* 180:119–52.

Page, G., and D. F. Whitacre. 1975. Raptor predation on wintering shorebirds. *Condor* 77:73–83.

Savage, C. 1992. *Peregrine falcons.* San Francisco: Sierra Club Books.

Taylor, I. 1994. *Barn owls.* Cambridge: Cambridge University Press.

Videler, J. J., and A. Groenewold. 1991. Field measurements of hanging flight aerodynamics in the kestrel, *Falco tinnunculus*. *Journal of Experimental Biology* 155:519–30.

Videler, J. J., D. Weihs, and S. Daan. 1983. Intermittent gliding in the hunting flight of the kestrel, *Falco tinnunculus* L. *Journal of Experimental Biology* 102:1–12.

Viitala, J., E. Korpimaki, P. Palokangas, and M. Koivula. 1995. Attraction of kestrels to vole scent marks visible in ultraviolet light. *Nature* 373:425–27.

Walter, H. 1979. *Eleonora's falcon*. Chicago: University of Chicago Press.

Watson, D. 1977. *The hen harrier*. Berkhamstead, England: T. and A. D. Poyser.

Yack, J. E., and J. H. Fullard. 1993. What is an insect ear? *Annals of the Entomological Society of America* 86:677–82.

Chapter 6. The Scorpion's Sting

Andrade, M. C. B. 1996. Sexual selection for male sacrifice in the Australian redback spider. *Science* 271:70–72.

Benton, T. G. 1992. The ecology of the scorpion *Euscorpius flavicaudis* in England. *Journal of Zoology, London* 226:351–68.

Cloudsley-Thompson, J. L. 1958. *Spiders, scorpions, centipedes, and mites*. New York: Pergamon Press.

Craig, C. L., R. S. Weber, and G. D. Bernard. 1996. Evolution of predator-prey systems: Spider foraging plasticity in response to the visual ecology of prey. *American Naturalist* 147:205–29.

Eberhard, W. G. 1977. Aggressive chemical mimicry by a Bolas spider. *Science* 198:1173–75.

Enders, F. 1975. The influence of hunting manner on prey size, particularly in spiders with long attack distances (Araneidae, Linyphiidae, and Salticidae). *American Naturalist* 109:737–63.

Foelix, R. F. 1982. *Biology of spiders*. Cambridge: Harvard University Press.

Forster, L. 1982. Vision and prey-catching strategies in jumping spiders. *American Scientist* 70:165–75.

———. 1995. The behavioural ecology of *Latrodectes hasselti* (Thorell), the Australian redback spider (Araneae: Theridiiae): A review. *Records of the Western Australian Museum* 52:13–24.

———. and R. R. Forster. 1985. A derivative of the orb web and its evolutionary significance. *New Zealand Journal of Zoology* 12:455–65.

Freed, A. N. 1984. Foraging behaviour in the jumping spider *Phidippus audax:* Bases for selectivity. *Journal of Zoology, London* 203:49–61.

Hjelle, J. T. 1990. Anatomy and morphology. In *The biology of scorpions*, edited by G. A. Polis, pp. 9–63. Stanford: Stanford University Press.

Hölldobler, B., and E. O. Wilson. 1990. *Ants*. Cambridge: Belknap Press.

Jackson, R. R. 1992a. Predator-prey interactions between web-invading jumping spiders and two species of tropical web-building pholcid spiders, *Psilochorus sphaeroides* and *Smeringopus pallidus*. *Journal of Zoology, London* 227:531–36.

———. 1992b. Predator-prey interactions between web-invading jumping spiders and *Argiope appensa* (Araneae, Araneidae), a tropical orb-weaving spider. *Journal of Zoology, London* 228:509–20.

———. 1992c. Predator-prey interactions between web-invading jumping spiders and a web-building spider, *Holocnemus pluchei* (Araneae, Pholcidae). *Journal of Zoology, London* 228:589–94.

Jackson, R. R., and R. S. Wilcox. 1990. Aggressive mimicry, prey-specific predatory behaviour, and predator-recognition in the predator-prey interactions of *Portia fimbriata* and *Euryattus* sp., jumping spiders from Queensland. *Behavioral Ecology and Sociobiology* 26:111–19.

McCormick, S. J., and G. A. Polis. 1982. Arthropods that prey on vertebrates. *Biological Review* 59:29–58.

———. 1990. Prey, predators, and parasites. In *The biology of scorpions,* edited by G. A. Polis, pp. 294–320. Stanford: Stanford University Press.

Orr, M. R., S. H. Seike, W. W. Benson, and L. E. Gilbert. 1995. Flies supress fire ants. *Nature* 373:292–93.

Polis, G. A., and S. J. McCormick. 1986. Scorpions, spiders, and solpugids: Predation and competition among distantly related taxa. *Oecologia* 71:111–16.

Polis, G. A., and W. D. Sissom. 1990. Life history. In *The biology of scorpions,* edited by G. A. Polis, pp. 161–223. Stanford: Stanford University Press.

Richert, S. E. 1992. Spiders as representative "sit-and-wait" predators. In *Natural enemies: The population biology of predators, parasites, and diseases,* edited by M. J. Crawley, pp. 313–28. Oxford: Blackwell.

Sabelis, M. W. 1992. Predatory arthropods. In *Natural enemies: The population biology of predators, parasites, and diseases,* edited by M. J. Crawley, pp. 225–64. Oxford: Blackwell.

Simard, J. M., and D. D. Watt. 1990. Venoms and toxins. In *The biology of scorpions,* edited by G. A. Polis, pp. 414–44. Stanford: Stanford University Press.

Simmons, A. H., C. A. Michal, and L. W. Jelinski. 1996. Molecular orientation and two-component nature of the crystalline fraction of spider dragline silk. *Science* 271:84–87.

Spradbery, J. P. 1973. *Wasps. An account of the biology and natural history of solitary and social wasps.* Seattle: University of Washington Press.

Tirrell, D. A. 1996. Putting a new spin on spider silk. *Science* 271:39–40.

Wehner, R., A. C. Marsh, and S. Wehner. 1992. Desert ants on a thermal tightrope. *Nature* 357:586–87.

Chapter 7. Saurian Warriors

Alexander, R. McN. 1985. Mechanics of posture and gait of some large dinosaurs. *Zoological Journal of the Linnean Society* 83:1–25.

Bakker, R. T. 1980. Dinosaur heresy—dinosaur renaissance; why we need endothermic archosaurs for a comprehensive theory of bioenergetic evolution. In *A cold look at the warm-blooded dinosaurs,* edited by R. D. K. Thomas and E. C. Olsen, pp. 351–462. Boulder: Westview Press.

———. 1986. *The Dinosaur heresies.* New York: William Morrow.

Bennett, S. C. 1992. Sexual dimorphism of *Pteranodon* and other pterosaurs, with comments on cranial crests. *Journal of Vertebrate Paleontology* 12:422–34.

Bramwell, C. D., and G. R. Whitfield. 1974. Biomechanics of *Pteranodon. Philosophical Transactions of the Royal Society of London* 267:503–81.

Brown, B. 1943. Flying reptiles. *Natural History* 52:104–11.

Charlesworth, E. 1840. Hawkins sale to the British Museum and his criminal libel case against Charlesworth. *Annual Magazine of Natural History* 4:11–44 (appendix).

Chiappe, L. M., and A. Chinsamy. 1996, *Pterodaustro*'s true teeth. *Nature* 379:211–12.

Colbert, E. H. 1962. The weights of dinosaurs. *American Museum Novitates* 2076:1–16.

Coria, R. A., and L. Salgado. 1995. A new giant carnivorous dinosaur from the Cretaceous of Patagonia. *Nature* 377:224–26.

Currie, P. J. 1995. New information on the anatomy and relationships of *Dromaeosaurus albertensis* (Dinosauria: Theropoda). *Journal of Vertebrate Paleontology* 15:576–91.

Delair, J. B. 1969. A history of the early discoveries of Liassic ichthyosaurs in Dorset and Somerset (1779–1835) and the first record of the occurrence of ichthyosaurs in the Purbeck. *Proceedings of the Dorset Natural History and Archaeological Society* 90:115–32.

Holtz, T. R. 1994. The phylogenetic position of the Tyrannosauridae: Implications for theropod systematics. *Journal of Paleontology* 68:1100–17.

Home, E. 1814. Some account of the fossil remains of an animal more nearly allied to fishes than any other classes of animals. *Philosophical Transactions of the Royal Society of London* 101:571–76.

Lang, W. D. 1959. Mary Anning's escape from lightning. *Proceedings of the Dorset Natural History and Archaeological Society* 80:91–93.

Langbauer, W. R., K. B. Payne, R. A. Charif, L. Rapaport, and F. Osborn. 1991. African elephants respond to distant playbacks of low-frequency conspecific calls. *Journal of Zoology, London* 157:35–46.

Mackay, R. S. 1964. Galápagos tortoise and marine iguana deep body temperatures measured by radio telemetry. *Nature* 204:355–58.

McGowan, C. 1979. Selection pressures for high body temperatures: Implications for dinosaurs. *Paleobiology* 5:285–95.

———. 1991. *Dinosaurs, spitfires, and sea dragons*. Cambridge: Harvard University Press.

———. 1992. The ichthyosaurian tail: Sharks do not provide an appropriate analogue. *Palaeontology* 35:555–70.

———. 1996. Giant ichthyosaurs of the Early Jurassic. *Canadian Journal of Earth Science* 33:1011–21.

Norman, D. 1985. *The illustrated encyclopedia of dinosaurs*. London: Salamander.

Ostrom, J. H. 1969. Osteology of *Deinonychus antirrhopus*, an unusual theropod from the Lower Cretaceous of Montana. *Peabody Museum of Natural History, Yale University Bulletin* 30:1–165.

———. 1990. Dromaeosauridae. In *The Dinosauria*, edited by D. B. Weishampel, P. Dodson, and H. Osmólska, pp. 1269–79. Berkeley: University of California Press.

Owen, R. 1840. Notice on the dislocation of the tail at a certain point observable in the skeleton of many ichthyosauri. *Transactions of the Geological Society of London* 5:511–14.

Padian, K., and J. M. V. Rayner. 1993. The wings of pterosaurs. *American Journal of Science* 293-A:91–166.

Ruben, J. 1995. The evolution of endothermy in mammals and birds: From physiology to fossils. *Annual Review of Physiology* 57:69–95.

Taylor, J. W. R. 1975. *Civil airliner recognition*. London: Ian Allan.

Watson, L. 1981. *Sea guide to whales of the world*. Toronto: Nelson.

Weishampel, D. B. 1981a. Acoustic analyses of potential vocalization in lambeosaurine dinosaurs (Reptilia: Ornithischia). *Paleobiology* 7:252–61.

———. 1981b. The nasal cavity of lambeosaurian hadrosaursids (Reptilia: Ornithischia): Comparative anatomy and homologies. *Journal of Paleontology* 55:1046–57.

Wellnhofer, P. 1991. *The illustrated encyclopedia of pterosaurs.* London: Salamander.

Williston, S. W. 1902. Winged reptiles. *Popular Science Monthly* 60:314–22.

Young, G. 1819. Account of a singular fossil skeleton, discovered at Whitby, in February 1819. *Memoirs of the Wernernian Natural History Society, Edinburgh* 3:450–57.

Chapter 8. Warfare in Miniature

Alcaraz, M., G-A. Paffenhöfer, and J. R. Strickler. 1980. Catching the algae: A first account of visual observations of filter-feeding calanoids. In *Evolution and ecology of zooplankton communities,* edited by W. C. Kerfoot, pp. 241–48. Hanover, N.H.: University Press of New England.

Buskey, E. J. 1984. Swimming pattern as an indicator of the roles of copepod sensory systems in the recognition of food. *Marine Biology* 79:165–75.

Buskey, E. J., L. Mills, and E. Swift. 1983. The effects of dinoflagellate bioluminescence on swimming behavior of marine copepoda. *Limnology and Oceanography* 28:575–79.

Dacey, J. W. H., and S. G. Wakeham. 1986. Oceanic dimethylsulfide: Production during zooplankton grazing on phytoplankton. *Science* 233:1314–16.

De Moth, W. R., and M. D. Watson. 1991. Remote detection of algae by copepods: Responses to algal size, odors, and motility. *Journal of Plankton Research* 13:1203–22.

Esaias, W. E., and H. C. Curl. 1972. Effect of dinoflagellate bioluminescence on copepod ingestion rates. *Limnology and Oceanography* 17:901–6.

Feigenbaum, D., and M. R. Reeve. 1977. Prey detection in the Chaetognatha: Response to a vibrating probe and experimental determination of attack distance in large aquaria. *Limnology and Oceanography* 22:1052–58.

Gallager, S. M. 1993. Hydrodynamic disturbances produced by small zooplankton: Case study for the veliger larva of a bivalve mollusc. *Journal of Plankton Research* 15:1277–96.

Habermehl, G. G. 1981. *Venomous animals and their toxins.* Berlin: Springer-Verlag.

Hays, G. C. 1995. Zooplankton avoidance activity. *Nature* 376:650.

Kerfoot, W. C. 1978. Combat between predatory copepoda and their prey: *Cyclops, Epischura,* and *Bosmina. Limnology and Oceanography* 23:1089–1102.

Kerfoot, W. C., D. L. Kellogg, Jr., and J. R. Strickler. 1980. Visual observations of live zooplankters: Evasion, escape, and chemical defenses. In *Evolution and ecology of zooplankton communities,* edited by W. C. Kerfoot, pp. 10–27. Hanover, N.H.: University Press of New England.

Koehl, M. A. R., and J. R. Strickler. 1981. Copepod feeding currents: Food capture at low Reynolds number. *Limnology and Oceanography,* 26: 1062–1073.

Kuhlmann, H-W., and K. Heckmann. 1985. Interspecific morphogens regulating prey-predator relationships in Protozoa. *Science* 227:1347–49.

Landry, M. R. 1980. Detection of prey by *Calanus pacificus:* Implications of the first antennae. *Limnology and Oceanography* 25:545–49.

Landry, M. R., J. M. Lehner-Fournier, and V. L. Fagerness. 1985. Predatory feeding behavior of the marine cyclopoid copepod *Corycaeus anglicus. Marine Biology* 85:163–69.

Nevitt, G. A., R. R. Velt, and P. Kareiva. 1995. Dimethyl sulphide as a foraging cue for Antarctic Procellariform seabirds. *Nature* 376:680–82.

Newell, G. E., and R. C. Newell. 1963. *Marine plankton*. London: Hutchinson.

Porter, K. G., and J. W. Porter. 1979. Bioluminescence in marine plankton: A coevolved antipredatory system. *American Naturalist* 114:458–61.

Poulet, S. A., and P. Marsot. 1980. Chemosensory feeding and food-gathering by omnivorous marine copepoda. In *Evolution and ecology of zooplankton communities,* edited by W. C. Kerfoot, pp. 198–218. Hanover, N.H.: University Press of New England.

Price, H. J., G.-A. Paffenhöfer, and J. R. Strickler. 1983. Modes of cell capture in calanoid copepods. *Limnology and Oceanography* 28:116–23.

Reeve, M. R. 1964. Feeding of zooplankton, with special reference to some experiments with *Sagitta*. *Nature* 201:211–13.

Strickler, J. R. 1984. Sticky water: A selective force in copepod evolution. In *Trophic Interactions Within Aquatic Ecosystems,* edited by D. G. Meyers and J. R. Strickler, pp. 187–239. Boulder: Westview Press.

Verni, F. 1985. *Litonotus-Euplotes* (predator-prey) interactions: Ciliary structure modifications of the prey caused by toxicysts of the predator (Protozoa, Ciliata). *Zoomorphology* 105:333–35.

Williamson, C. E., and H. A. Vanderploeg. 1988. Predatory suspension-feeding in *Diaptomus:* prey defenses and the avoidance of cannibalism. *Bulletin of Marine Science* 43:561–72.

Wong, C. K., C. W. Ramcharan, and W. G. Sprules. 1986. Behavioral responses of a herbivorous calanoid copepod to the presence of other zooplankton. *Canadian Journal of Zoology* 64:1422–25.

Yen, J. 1988. Directionality and swimming speeds in predator-prey and male-female interactions of *Euchaeta rimana,* a subtropical marine copepod. *Bulletin of Marine Science* 43:395–403.

Chapter 9. The Interminable Struggle

Baldwin, I. T. 1988. Short-term damage-induced increases in tobacco alkaloids protect plants. *Oecologia* 75:367–70.

Bellamy, D., and A. Pfister. 1992. *World medicine: Plants, patients, and people.* Oxford: Blackwell.

Boppré, M. 1986. Insect pharmacophagously utilizing defensive plant chemicals (pyrrolizidine alkaloids). *Naturwissenschaften* 73:17–26.

Brown, K. S., Jr. 1984. Adult-obtained pyrrolizidine alkaloids defend ithomiine butterflies against a spider predator. *Nature* 309:707–9.

Bryant, J. P., F. D. Provenza, J. Pastor, P. B. Reichardt, T. P. Clausen, and J. T. du Toit. 1991. Interactions between woody plants and browsing mammals mediated by secondary metabolites. *Annual Review of Ecology and Systematics* 22:431–46.

Coley, P. D., J. P. Bryant, and F. S. Chapin III. 1985. Resource availability and plant antiherbivore defense. *Science* 230:895–99.

Dussourd, D. E., C. A. Harvis, J. Meinwald, and T. Eisner. 1989. Paternal allocation of sequestered plant pyrrolizidine alkaloid to eggs in the danaine butterfly, *Danaus gilippus*. *Experimentia* 45:896–98.

Dussourd, D. E., K. Ubik, C. Harvis, J. Resch, J. Meinwald, and T. Eisner. 1988. Biparental defensive endowment of eggs with acquired plant alkaloid in the moth *Utetheisa ornatrix. Proceedings of the National Academy of Science* 85:5992–96.

Eastop, V. F. 1981. Coevolution of plants and insects. In *The evolving biosphere,* edited by P. H. Greenwood and P. L. Forey, pp. 179–90. Cambridge: Cambridge University Press.

Futuyma, D. J. 1983. Evolutionary interactions among herbivorous insects and plants. In *Coevolution,* edited by D. J. Futuyma and M. Slatkin, pp. 207–31. Sunderland, Mass.: Sinauer.

Gibson, R. W., and J. A. Pickett. 1983. Wild potato repels aphids by release of aphid alarm pheromone. *Nature* 302:608–9.

Hölldobler, B., and E. O. Wilson. 1990. *Ants.* Cambridge: Belknap Press.

Janis, C. M., and M. Fortelius. 1988. On the means whereby mammals achieve increased functional durability of their dentitions, with special reference to limiting factors. *Biological Review* 63:197–230.

Joyce, C. 1994. *Earthly goods: Medicine-hunting in the rainforest.* Boston: Little, Brown.

Kingsbury, J. M. 1964. *Poisonous plants of the United States and Canada.* Englewood Cliffs, N.J.: Prentice-Hall.

Lampe, K. F., and M. A. McCann. 1985. *AMA Handbook of Poisonous and Injurious Plants.* Chicago: American Medical Association.

Levin, D. A. 1973. The role of trichomes in plant defense. *Quarterly Review of Biology* 48:3–15.

Lindroth, R. L., G. O. Batzli, and S. I. Avildsen. 1986. *Lespedeza* phenolics and *Penstemon* alkaloids: Effects on digestion efficiencies and growth of voles. *Journal of Chemical Ecology* 12:713–28.

Muenscher, W. C. 1951. *Poisonous plants of the United States.* New York: Macmillan.

Newton, P. N., and T. Nishida. 1991. Possible buccal administration of herbal drugs by wild chimpanzees, *Pan troglodytes. Animal Behaviour* 39:798–801.

Pollard, A. J. 1992. The importance of deterrence: Responses of grazing animals to plant variation. In *Plant resistance to herbivores and pathogens,* edited by R. S. Fritz and E. L. Simms, pp. 216–39. Chicago: University of Chicago Press.

Rothschild, M., R. T. Aplin, P. A. Cockrum, J. A. Edgar, P. Fairweather, and R. Lees. 1979. Pyrrolizidine alkaloids in arctiid moths (Lep.) with a discussion on host plant relationships and the role of these secondary plant substances in the Arctiidae. *Biological Journal of the Linnean Society* 12:305–26.

Rubinstein, B. 1992. Similarities between plants and animals for avoiding predation and disease. *Physiological Zoology* 65:473–92.

Schneider, D., M. Boppré, J. Zweig, S. B. Horsley, T. W. Bell, J. Meinwald, and K. Hansen. 1982. Regulation by pyrrolizidine alkaloids. *Science* 215:1264–65.

Van Loon, L. C. 1993. Plant defenses. In *Defence mechanisms,* pp. 77–142. Biotechnology by Open Learning series. Oxford: Butterworth-Heinemann.

Vrieling, K., W. Smit, and E. van der Meijden. 1991. Tritrophic interactions between aphids (*Aphis jacobaeae* Shrank), ant species, *Tyria jacobaeae* L., and *Senecio jacobaea* L. lead to maintenance of genetic variation in pyrrolizidine alkaloid concentration. *Oecologia* 86:177–82.

Waller, G. R., and E. K. Nowacki. 1978. *Alkaloid biology and metabolism in plants.* New York: Plenum Press.

Wink, M., and P. Römer. 1986. Acquired toxicity—the advantages of specializing on alkaloid-rich lupins to *Macrosiphon albifrons* (Alphidae). *Naturwissenschaften* 73:210–12.

Winter, H., A. A. Seawright, J. Hrdlicka, U. Tshewang, and B. J. Gurung. 1992. Pyrrolizidine alkaloid poisoning of yaks (*Bos grunniens*) and confirmation by recovery of pyrrolic metabolites from formalin-fixed liver tissue. *Research in Veterinary Science* 52:187–94.

Winter, H., A. A. Seawright, A. R. Mattocks, R. Jukes, U. Tshewang, and B. J. Gurung. 1990. Pyrrolizidine alkaloid poisoning in yaks. First report and confirmation by identification of sulphurbound pyrrolic metabolites of the alkaloids in preserved liver tissue. *Australian Veterinary Journal* 67:411–12.

Winter, H., A. A. Seawright, H. J. Noltie, A. R. Mattocks, R. Jukes, Kinzang Wangdi, and J. B. Gurung. 1994. Pyrrolizidine alkaloid poisoning of yaks: Identification of the plants involved. *Veterinary Record* 134:135–39.

Zucker, W. V. 1983. Tannins: Does structure determine function? An ecological perspective. *American Naturalist* 121:335–65.

Index